Linking Research
to Practice

The **Singapore Internet Research Centre (SiRC)** initiates and conducts research related to new media/internet across Asia, including East, Southeast, and South Asia. The centre's mission is to bring Asian experiences and perspectives to the global discussion about the innovation, development and impact of the internet and information technologies.

Canada's **International Development Research Centre (IDRC)** is one of the world's leading institutions in the generation and application of new knowledge to meet the challenges of international development. Since 1970, IDRC has worked in close collaboration with researchers from the developing world as they strive to build healthier, more equitable, and more prosperous societies

The **Institute of Southeast Asian Studies (ISEAS)** was established as an autonomous organization in 1968. It is a regional centre dedicated to the study of socio-political, security and economic trends and developments in Southeast Asia and its wider geostrategic and economic environment. The Institute's research programmes are the Regional Economic Studies (RES, including ASEAN and APEC), Regional Strategic and Political Studies (RSPS), and Regional Social and Cultural Studies (RSCS).

ISEAS Publishing, an established academic press, has issued more than 2,000 books and journals. It is the largest scholarly publisher of research about Southeast Asia from within the region. ISEAS Publishing works with many other academic and trade publishers and distributors to disseminate important research and analyses from and about Southeast Asia to the rest of the world.

Linking Research to Practice

Strengthening ICT for Development Research Capacity in Asia

EDITED BY

Arul Chib • Roger Harris

SINGAPORE INTERNET RESEARCH CENTRE

INTERNATIONAL DEVELOPMENT RESEARCH CENTRE

Ottawa • Cairo • Dakar • Montevideo • Nairobi • New Delhi • Singapore

INSTITUTE OF SOUTHEAST ASIAN STUDIES

Singapore

First published jointly in Singapore in 2012 by

ISEAS Publishing
Institute of Southeast Asian Studies
30 Heng Mui Keng Terrace, Pasir Panjang
Singapore 119614
E-mail: publish@iseas.edu.sg • *Website:* bookshop.iseas.edu.sg

Singapore Internet Research Centre
Wee Kim Wee School of Communication and Information
Nanyang Technological University
31 Nanyang Link
Singapore 637718
Website: www.sirc.ntu.edu.sg

International Development Research Centre
PO Box 8500
Ottawa, ON K1G 3H9
Canada
E-mail: info@idrc.ca • *Website:* www.idrc.ca

The responsibility for facts and opinions in this publication rests exclusively with the authors and their interpretations do not necessarily reflect the views or the policy of ISEAS, SiRC, IDRC or their supporters.

ISEAS Library Cataloguing-in-Publication Data

Linking research to practice : strengthening ICT for development research capacity in Asia / edited by Arul Chib and Roger Harris.
   1. Economic development—Study and teaching—Asia.
   2. Information technology—Asia.
   3. Telecommunication—Asia.
   I. Chib, Arul.
   II. Harris, Roger.
   III. Title: Strengthening ICT for development research capacity in Asia
HC415 I55L75   2012

ISBN 978-981-4380-00-3 (soft cover)
ISBN 978-981-4380-01-0 (e-book, PDF)

Cover photo: Lao lady with laptop. Reproduced with kind permission of Roger Harris.

Typeset by Superskill Graphics Pte Ltd
Printed in Singapore by Markono Print Media Pte Ltd

CONTENTS

SECTION II: RESEARCH PERSPECTIVES: THEORETICAL REFLECTIONS BY EXPERTS

SECTION III: RESEARCH OUTPUTS

SECTION IV: SYNTHESIS AND CONCLUSION

LIST OF TABLES

LIST OF FIGURES

FOREWORD

If, as some suggest, a messy desk is a sign of a creative mind, perhaps a messy conference room is the sign of a creative collaboration. Over the years, I have spent a lot of time in workshops at the Microsoft Research (MSR) offices in Bangalore, but for some reason, I still remember the way the attendees at the ICRC research methods workshop appropriated the MSR space. They — actually 'we', since I was a participant — moved the neat rows of desks and tables around into circles and clusters, spilled out into the offices' common spaces, and set out to question nearly everything about the practice of "Information and Communication Technologies and Development" (ICTD). Given the broad representation of accomplished scholars, practitioners, and donors at the early planning sessions in Manila and Bangalore, one might have expected that the project which would become SIRCA would be special. But I do think it was a good sign that we moved the chairs.

Another thing I remember from those early meetings was how helpful it was to have people from outside the ICTD community involved in the discussions. Some of the most insightful conversations during the sessions occurred as self-described ICTD researchers compared their articulations of problems, methods, and interests to those attendees interested more in research on the globalizing "information society". The result was a group dynamic which paid unusually careful attention to assumptions, theoretical stances, and goals. No one was going to get away with a blanket assertion that worldwide access to and use of information and communications was fair, good, or necessary, or that interventions to 'improve' access and use were required, effective, or even possible.

A few years later, having now seen the fruition of a full cycle of SIRCA work as represented in this volume, it is extremely gratifying to see how the programme has not only supported fascinating new interdisciplinary research on ICTD and on the information society in emerging Asia, but also strengthened and nurtured a growing group of scholars in the region. Some of the relationships and intellectual cross-pollinations between mentors and researchers, and between researchers from different sub-regions, will surely last beyond the cycle of the SIRCA project. In this way, the endeavour has already contributed to the research community in a way that is different from a conventional call for papers or one-off workshop.

A more surprising output from the first cycle of SIRCA research relationships has been the burst of new and rigorous reflections it has offered on the practice of doing ICTD research. These reflections offer not simply refinements to surface methods, but rather critical discussion of praxis, of ethics, of teaching, of stakeholder coordination, of publication, of institutional identity, and the purpose of ICTD research itself. In this way, the topics discussed in the SIRCA workshops and in this volume offer new insights into issues, challenges, and the state of the art in the ICTD field as a whole.

Perhaps, whether by design or by fortune, the commitment to putting mentorship and capacity building at the centre of the SIRCA initiative was a major contributor to this burst of reflection. Mentorship is, of course, a time-tested way to encourage the transfer of tacit knowledge from one generation to the next. But in this case, it seems that the authors, participants, and editors were able to capture (for our benefit) a lot of that tacit knowledge, mid-transfer. This volume represents the reflections of a variety of scholars who were actively and personally engaged in an ongoing discussion not simply on the rote methods of ICTD (sample sizes and citations and such), but rather on its deeper craft, and on the rituals and requirements of a complex and growing community of practice. In other words, the experienced mentors were learning as well, since the process encouraged them to challenge their assumptions, take positions, and in some cases develop new ideas and research syntheses.

Thus the SIRCA project was distinctive in many ways: It involved researchers, practitioners, and donors right from the start. It put a broad set of methodological and theoretical perspectives on the table for discussion and integration. And, through mentorship, it sought to make tacit knowledge about the practice of ICTD more explicit and accessible. Thanks to this design by IDRC and the SIRCA team, and to a significant multi-year commitment on the part of the funders and the coordinators, the outputs of the project and of this volume include the following:

SIRCA was an opportunity to pursue impact, but without assumptions. You may note a few instances in the text where authors politely declined to take a detour into the semantics, particularly the differences between "ICTD" and "ICT4D". This is not because these distinctions weren't relevant or discussed, but rather because there was both an awareness of the power of the underlying assumptions conveyed by nomenclature, and also an understanding that the community enjoyed the shelter of a big enough tent to include those who may strongly prefer one term over another. What was foregrounded instead was the importance of *impact.* Some framed this in terms of the Millennium Development Goals, others simply as "development", but the chapters spoke from a consistent perspective about the need to engage with society as a whole to make people's lives better. Again in the spirit of the "big tent", it is unlikely that all the authors agree precisely on the mechanisms through which scholarly research is translated into change in the world; instead the volume is a reflection on the various mechanisms which might prove fruitful. The only stances absent from this volume's overall orientation towards "impact" (4impact?) might be more critical ones — perspectives which might directly challenge the possibility of an inclusive information society, or stress the relationships between information technologies and entrenched power.

SIRCA supported research that pushed up against the boundaries of what might traditionally be considered ICTD research. Of course, the programme generated papers and presentations beyond those represented in this volume, most notably in the journal *Media Asia.* But even the research outputs highlighted in this volume reflect a variety of national contexts, technological artefacts, development domains and theoretical approaches. By exploring political blogging, environmental sustainability and community level coordination, the challenges of mobile learning, and the engagement of individual university students with the full range of internet options (instead of focusing only on pro-development, instrumental activities), the research output papers reflect state-of-the-art approaches to understanding the information society as it manifests in emerging Asia. The processes are framed neither exclusively as *drivers* of development nor exclusively as *results* of development. Instead, the inquiries stress context and interdependencies with other changes in the social and economic environments of the countries in question.

SIRCA advanced the state-of-the-art in ICTD without being constrained within it. Across the papers, the complex practice of ICTD was linked latterly and sequentially with adjacent communities of practice and research. Flor's chapter provides a valuable reminder of the debt ICTD research owes to antecedents in development communication, information economics, and knowledge management (among others). Each is still a thriving field and thus Flor's first chapter can also remind us of how stimulating it can be to learn from one's neighbours. Meanwhile, Flor's second chapter with Harris approaches the current question of institutional affiliations and linkages. Conceptual and relativistic flexibility in the fluid domain of ideas is all well and good, but the challenges of learning, conducting, and teaching an inherently interdisciplinary subject in the environment of traditional university departments remains significant for ICTD. Similarly Traxler's piece on the challenges of implementing a mobile health programme in Cambodia is engaged with the theoretical and methodological issues brought up earlier and throughout the volume. What are "emerging researchers" to make of multiple, intersecting discourses around development and technology? Such matters can only be confounded when ICTD research confronts a web of intersecting stakeholder demands as illustrated by Chib, Ale and Lim. The volume reminds us that many of the important actors in this drama, from donors to governments to, especially, individual users and families, care little what we call our field, where we publish, or which department we sit in.

Years after the initial meetings in Manila and Bangalore, this volume showcases the fruits of interactions between a set of established researchers with a broader community of new researchers. In his chapter on ICTD methods, De' invokes the idea of a "clean room" in a semiconductor factory as the idealized but never attainable state for research. Instead, De argues, it is via "messy methods" that the best and most applicable ICTD research is accomplished. And yet, De' observes that messy methods are hard; since they demand "a greater degree of preparation and sensitization to enable the researcher to respond to the needs of the situation on the ground". It in this spirit that the entire volume, not just the specific methods chapters, is of service to ICTD. The mess outside the clean room which De' uses to talk about methods is something which seems to have permeated beyond the methods chapter, and into the whole report. This dialogue, this endeavour around mentorship, building capacity, creating a research culture of ICTD in Asia seems to have created a wonderfully generative and simulative environment for the experts writing in Section II, and for the researchers sharing firsthand reports in Section III.

In aggregate, the volume offers a unique perspective (impact beyond the D), via a unique approach and common thread (mentorship), in a unique domain (the information society in emerging Asia). To write a foreword to such a volume, itself already deeply reflective and integrative, risks introducing a Mobius strip of reflection-on-reflection. So I'll end here before the next recursive twist. Rest assured that the insights on the following pages are fresh, deeply and visibly connected to the multiyear project and the mentor relationships nurtured by IDRC and SIRCA, and extremely helpful for ICTD researchers worldwide.

This is what can happen when the desks are shuffled around; when a 'messy' conference room is indeed an early indication of a generative, creative process.

Jonathan Donner
Microsoft Research
January 2012

PREFACE

The Singapore Internet Research Centre (SiRC) has championed the cause of ICT for Development (ICTD) research with Strengthening ICTD Research Capacity in Asia (SIRCA), a pioneer capacity-building programme that aims to develop the research skills of emerging researchers in the Asia Pacific region. Under SIRCA, a number of experienced scholars served as mentors to principal investigators from all over Asia. Their collaboration has resulted in a range of research findings and lessons learnt which are compiled in this volume. It is divided into three sections, or perspectives: (1) Management Perspectives; (2) Research Perspectives; and (3) Research Outputs.

The first section discusses the inception of the SIRCA programme; opportunities, issues and concerns arising from the implementation, management and evaluation of both the process and the outcomes. The section primarily underscores the need for ICTD research to be more analytical and empirical as opposed to descriptive and anecdotal. It also highlights the importance of rigour and sophistication in ICTD research methodology. The narratives on the management level highlight the relationship of the mentors and PIs, the lessons learnt, and how the programme inspired the PIs to engage in ICT research and practice. The evaluation of the programme describes SIRCA's success in achieving its objectives, as well as recommendations for improvement.

The second section reflects the range of academic traditions represented in the programme through the lens of the research mentors. The section highlights the importance and complexity of key links; that between ICTD theory and praxis, and the links between research, instruction and professional development. However, the practice of ICTD has not yet yielded a straight pathway to theorising either. We often aim to conduct ICTD research inside a "clean room" of research methods — structured, organised, and systematic — yet the reality is one of messy encounters with data gathering and analysis. An overview of ethics in ICTD research suggests a comprehensive canon is absent. Addressing these theoretical, methodological, and ethical issues from a critical perspective goes hand-in-hand with the hope of developing a unique tradition within ICTD research. Finally, we champion the importance of an ICTD curriculum that will not only train future researchers and academics but will also sustain the practice, whilst providing inclusion to research teams situated on the margins of the discipline.

The third and final section highlights research papers produced by the SIRCA programme, beyond those that have found a home elsewhere in peer-reviewed global journals. With studies conducted in Vietnam, Cambodia and the Philippines, the exemplars provide a concrete picture of the processes the principal investigators have undergone. As a SIRCA mentor stated, the programme has served as a platform for the cross-fertilisation of ideas. This collaboration has occurred across academic traditions, cultural backgrounds, and national and ethnic boundaries. SIRCA has led to a network of scholars that are bound not by schools of thought, ethnicities, age nor gender, but by a sense of purpose to use ICTs in the service of the Millennium Development Goals. In this sense, participation in the programme has brought higher order benefits to the principal investigators, mentors and staff. Based on the studies, narratives and recommendations provided by this volume, we offer insight into the extent of engagement required in the pursuit of disciplinary objectives, not merely for academic outcomes, but to contribute meaningfully to social change.

Arul CHIB
Roger HARRIS
Editors

ACKNOWLEDGEMENTS

We would like to acknowledge the following for their tremendous support to make this publication possible:

Jonathan Donner, for contextualising the work of SIRCA in the foreword;

The principal investigators: Regina Hechanova, Mary Grace Mirandilla-Santos, Pham Huu Ty, and Chivoin Peou, for inputs that would serve as excellent references for future scholars;

The research mentors: Alexander Flor, Rahul De', John Traxler, Richard Heeks, May O. Lwin, for their contributions and guidance to their respective principal investigators;

The International Development Research Centre, Canada: Laurent Elder, Chaitali Sinha, and Matthew Smith, for their contributions and support over the past three years;

The graduate reviewer: Rajiv Aricat, for meticulously going through the chapters;

The Institute of Southeast Asian Studies, for serving as our publisher;

The SIRCA Management team headed by Dr Ang Peng Hwa, who serves as director of SIRCA and the Singapore Internet Research Centre (SiRC); Yvonne Lim, Senior Manager, and Finance Manager, Liaw Wan Tieng; Tahani Iqbal, Komathi Ale, and Leandra Flor as the editorial team; as well as prior members of the SIRCA team; Joanna Tan, Grace Kwan, Naowarat Narula, and Sri Ranjini Mei Hua.

And our home, the Wee Kim Wee School of Communication and Information, Nanyang Technological University.

SECTION I

Management Perspectives: Insiders' Thoughts on the Programme

1

PERSPECTIVES ON ICTD RESEARCH AND PRACTICE

Roger Harris and Arul Chib

The role of information and communication technologies (ICTs) towards achieving development (ICTD; sometimes referred to as ICT4D) goals, such as education, gender empowerment, health, and poverty eradication has gained a fair bit of traction (United Nations 2005). ICTD is a general expression in which the term ICTs more often refers to new media technologies such as the internet, computer, mobile phone, global positioning system, but it can also mean more traditional media such as radio, television, and landline telephony. Development is generally regarded as the socio-economic progress of the developing world. There are many variations around these themes, as well as multiple labels describing them. It is not necessary to elaborate on the definitional distinctions here (there are other sources for that) because to do so would plunge us into the very trap for which this volume is intended as an aid in avoiding.

There are a number of high-profile academic studies encapsulating the promise of ICTD, such as rural fishermen in southern India, earning less than US$100 a month, using mobile phones to help decide to which of the several nearby markets to deliver their catch in order to get the best available price (Abraham 2007; Jensen 2007). Or the rural internet centre that saved lives by broadcasting news of the impending tsunami by loud-speaker to the local village community, saving them from drowning in a stormy ocean

(Subramanian 2005). Or mobile phones in Bangladesh that benefit entire villages to get connected, whilst providing economic value to the woman entrepreneurs that provide the service (Richardson, Ramirez, and Haq). Or of rural midwives in Indonesia that get connected to urban hospitals via JAVA-enabled mobile applications, improving healthcare services to pregnant village women (Chib 2010). These examples of the use of ICTs are spreading rapidly across the developing world; in some cases this is because of the actions of development practitioners and academics that take an interest in such things, but in many cases they are happening in spite of the actions of such people. Whilst we often hear of targeted interventions that aim to deliver specific benefits, there are potentially thousands, if not millions of cases that are arising organically, with the unabated diffusion of ICTs into the remotest reaches of the globe.

We have been, and continue to be, wearing multiple hats as academic researchers and as practitioners in ICTD (and interestingly enough, balance out each other's respective biases). We can therefore bear faithful witness to the wrangling that goes on in the world of both groups of professionals. Academic researchers argue (as they do in most fields, we suspect) over definitions and the supremacy of this or that theory as well as engaging in turf wars and power politics over who 'owns' the subject. The ICTD academic community comprises a host of scholars from a variety of disciplines, which then suggests the lack of a common body of scholarship, in theory, in method, or in the practice of research. Sociologists and technologists generally mix as well together as oil and water and academics have yet to come to terms with the fact that life's problems are not as conveniently organised as their faculty organisations.

Equally so, practitioners struggle with the 'newness' of ICTs, cutting across as they do, their stove-piped organisational boundaries and the orthodox methods within which they have comfortably fostered life-long careers. As the corporate world discovered decades ago, and as the socialising and revolutionising youth of today are finding out, ICTs demand fresh and integrated approaches to development for optimal results to emerge. Likewise, the business world quickly grasped the potential of ICTs, and invested heavily in their research and development, and in their dissemination, in order to achieve economies of scale viable for profitable business. On the other hand, ICTD practitioners remain unfortunately mired in a battle between those with a positivist stance and those who counter with "told-you-so" responses, such as the nay-sayers and inflexible traditionalists who fail to grasp the new opportunities that they offer and refuse to adapt to an ever-changing world.

The consequence of these research studies versus practice boundaries, as well as the internal struggles that continue, has been a plethora of case studies in ICTD, largely from a positivist angle. As theoretical rigour and methodological approaches remain fuzzy, practical intent, aided possibly by a donor-driven agenda, culminates in a series of descriptive cases with little by way of critical analysis (Heeks 2007). The resultant weaknesses in scientific credibility lead to challenges in influencing the key third prong of the research-practice triumvirate, i.e., policy.

There are notable exceptions that bridge the first divide; that between research and practice; including all who have made their contributions to this volume and the programme that led to it as well as the many other academics and practitioners with whom we are grateful for the opportunity to be able to continue working with. They are all technology champions; and tend to have the deepest connections to the communities within which they work. They have to be, because the champion is the person who has a vision, from which they have derived their mission and which in turn motivates them to make things happen. Sadly though, we observe that, even after many years in this field of work, it still requires this kind of dedication to engineer positive, longitudinal results that bring development benefits to the poor and marginalised. For the multitudinous NGOs and even more for poor individuals themselves, this is not an issue, but for the academic and development institutions involved in the process of making ICTs work for development, innate inhibitors continue to prevent them from bringing the full potential of their considerable resources in making ICTs achieve their full contribution.

Whilst the problems of identity and positioning in both the academic and practitioner worlds continue to plague the institutional response to ICTD, it is unlikely that the two will join together in a stable, mutually-reinforcing relationship that would be capable of generating substantial benefit for the developing world. However, such an unhappy situation need not be inevitable, provided concrete measures are promoted for bringing academics and their research out into the realm of practice, and vice versa. That's what this book is about. From the outset, we academics have to acknowledge that academic publishing is not enough to claim impact from our work (pardon the irony of saying this in an academic publication).

This may rankle with some academics who claim citations as impact, but by 'impact' here, we mean something that happens outside academia. We are talking, crucially, about a demonstrable contribution to society as opposed to the continual manipulation of ideas and theories; it is insufficient to consider only activities and outputs that might cultivate impact, such as a conference presentation, or a published report. Conceptual impact contributes to the understanding of policy issues, or reframes debates around them. Instrumental impact shapes the growth of policy, alters the provision of services, and moulds practice or influences legislation. The establishment of theories for prediction, of standards of measurement, and scientific evidence of the impact of ICTD need to be developed. Another dimension of impact, important in the current context, addressed by this book is the building of capacity within the field by sharing across multiple disciplines, crystallising of theoretical and methodological challenges, and laying out directions for future research endeavours. In all cases, impact needs to be planned for, identified, managed, measured and evaluated, none of which is straightforward, as causality is rarely linear and decisions are almost always based on multiple inputs.

Beyond the wide diffusion of ICTs within varied developmental contexts, there are instances where the practice of ICTD marches well ahead of the trailing curve of ICTD research. For example, global media coverage focused on the advent of commercially available low-cost computing devices such as the *One Laptop per Child* device, the *Intel Classmate PC*, and the recently announced USD$10 connectivity device by the Indian Government. National governments and transnational players such as the United Nations have seemingly paid more attention to their potential, possibly a result of the hype, than to rigorous scientific evidence. We offer a suggestion, then, that research needs to understand how to disseminate results to a broader audience, incorporating within its audience not only the scientific community, but also the mainstream public, as well as influential policy-makers.

Equally, as practitioners, we need to acknowledge the limitations of our conventional project-based methods, own up to what we don't know and ask questions to those who are capable of taking a more dispassionate and critical perspective of our work and who have received the professional training required to be capable of generating convincing results.

In our experience, there is often quite limited interaction between practitioners and academics in any of the disciplines that make up ICTD (of which there are several). Probably this is typical of other subject areas and therefore it may be naïve and/or unrealistic to expect that this should not be with ICTD, but it seems a pity. There are many lines of enquiry that could be usefully explored by those who are equipped with the skills that academics possess and which could usefully inform practice towards innovative ICT implementations that are more robust and effective in their pursuit of varied developmental goals.

Generally, practitioners in ICTD don't attend academic conferences because they know there will be little by way of any practical takeaways, whilst there would be plenty of scientific jargon that they would fail to understand.

They don't read academic journals for the same reason. Moreover, it seems rare to see any summarised form of the findings of academic research in the media channels that policy-makers and their advisers read. The more serious media that such people tend to take notice of, such as *The Economist* or the *Wall Street Journal*, seem to conduct their own research and they nearly always come up with rather superficial conclusions; like "telecentres are unsustainable" or "mobile phones have closed the digital divide" or "increasing the density of mobile phones boosts gross domestic product (GDP)." It needs an academic approach to dig beneath these superficial generalisations, to recapture the complexities and to help practitioners and policy-makers come to the more nuanced conclusions that are closer to reality.

One problematic area for practitioners has been to make their work sustainable in the long-run and generalisable to varied contexts. There are many aspects of orthodox development practice that do not lend themselves well to the application of ICTs to development problems. What often occurs is that the problem is shaped to suit the tools available for solving it, rather than the other way round. Positivism, in the form of techno-determinism, rears its ugly head here, potentially misrepresenting the potential of a particular technology, and diverting resources from more laudable projects. Possibly the most important areas where academic research can contribute to practice are in understanding individual motivations and barriers to usage, and impact assessment via project evaluation. Evaluation is one area where this stands out as an inhibitor of learning within the practitioner community. Instead, academia can develop investigations of beneficiary motivations, at psychological, individual, and social levels, and importantly, translate them to be valid and applicable to different contexts. These contexts could be in different geographical and cultural contexts, as well as scale to be applicable to national and international rollouts.

We should ask if the academic community is concerned that practitioners remain largely uninterested in their work. Perhaps they are not; it is clear that academics conduct research, attend conferences and publish their results in the leading journals in order to survive and flourish within their professional situations. In the closed-loop community of ICTD academic research, there's probably more effort that goes into counting impact-factors and citations than reaching out to the development professionals. Whilst we have organisations such as the *Asia-Pacific Telecommunity, the Centre for Science, Development and Media Studies, D.Net, LIRNEasia*, and conferences like *ICTD* and *CPRSouth*, there seem to be too few components that encourage the sort of practitioner-researcher linking that would contribute significantly to this process.

The biggest regret about this is that the developing world academics that we have encountered seem to have a genuine concern for how their work can contribute towards their country's progress. They have less access to resources for attending international conferences to highlight their research and network with their peers, but they are also under less pressure to publish in leading journals. Generally, budgets for conducting research are miniscule at developing country research institutions. The bilateral and multilateral donor organisations that have far greater access to funds tend to conduct their own research, ignoring the in-country academic resources that they could bring into their programmes and squandering the opportunity for learning that such involvement offers (in both directions).

ICTD practitioners need practical solutions to their pressing concerns, some of which are in the following areas: formulating technology policies; achieving sustainability; rolling out technology applications; scaling up from pilot projects to national programmes; conducting evaluations and impact assessments; promoting technology convergence and synergies; formulating pro-poor approaches; and devising demand-driven methods. Academic approaches to questions like these bring the benefit of rigour and independence to evaluation research enquiries, which is not always to be found in the sort of practitioner studies that confuse correlation with causation and which often generate rather predictable outcomes when conducted by the project implementers. In many developed economies, there is a move towards evidence-led policy formulation which implies greater reliance on robust research in order to generate reliable evidence. This alone constitutes a powerful reason for a closer relationship between researchers and practitioners and for a sharper focus on managing research in a manner that will increase the likelihood of having a demonstrable impact on society.

There are many ways of designing research that will have the potential to generate impact outside academic circles, i.e. by influencing policy and practice. At one end of the spectrum, there can be a highly structured methodological and participative approach and at the other end, one that relies on the experience of the researchers and/or practitioners conducting the research. In all cases, there are no guarantees that the outcomes will be as predicted, so some flexibility and continuous learning is always desirable.

In response, the Strengthening ICTD Research Capacity in Asia (SIRCA) programme had its genesis that could be categorised as being somewhere in the middle of the range of available approaches; a semi-structured process that tapped the knowledge of a team (grants review committee) of experienced researchers and practitioners. The SIRCA programme aimed to develop the skills of researchers in the Asia Pacific region on ICT and Development research. Under SIRCA, a number of experienced scholars served as mentors to principal investigators from all over Asia.

Accordingly, the projects that received funding, of which some are reported in this book, reflect the issues of both practice and research that the team identified at the time. However, in going forward, relying on a team of experts is not always the most effective strategy; individuals may not be available when required (there aren't that many of them around anyway) and subject experts are notorious for disagreeing amongst themselves. For these (and other) reasons, there is merit in applying a methodological approach to the design and implementation of research that will generate impact; an approach that can be taught to researchers and practitioners and which encapsulates the wisdom of the experts without demanding that they necessarily be present all the time.

The SIRCA programme is moving towards a roadmap for navigating the tortuous route from research to impact via practice. The starting point revealed the need to strengthen ICTD research in Asia as a precondition for promoting its potential for achieving impact. Impact is impossible with poor quality research. Through the process described here, including workshops and seminars, conferences, mentoring, and peer-reviewed academic publishing, the researchers were facilitated towards results that carry practical implications, and several of their products are included in this publication. They will continue with their research careers and the legacy of SIRCA will be their future contributions to advancing ICTs for Asia's social and economic development. As the programme proceeds into its next phase, the focus should shift towards a more methodological approach to achieving impact with research. Why should it do this?

In the first place, research funders are increasingly demanding impact-related relevance from the activities that they fund. Demonstrable impact enhances the value of research by making public services more effective and by improving the welfare of society and the economy. For the researchers, impact raises their profile and increases their influence, it makes their activities more relevant and it helps shape their research agenda and attract further funds. The SIRCA programme now has in place a management capability and network of researchers and practitioners that is highly suited to furthering the impact of ICTD research in Asia and can play an important role in finally bringing ICTD researchers and practitioners together in a mutually reinforcing and beneficial partnership, as well as showing other institutions how to achieve the same results.

The SIRCA collaboration has resulted in a range of research findings and lessons learnt compiled in this volume. It is divided into three sections, or perspectives; management perspective that outlines the genesis of the programme, key activities and outcomes, as well as evaluation of the performance in achieving objectives; research perspectives that address current critical issues, such as instruction, ethics, methods, that are of concern to the ICTD community; and finally, outcomes from the programme, representing research from various developing countries in Asia.

The first section addresses the role of institutions in capacity building; in order to develop inter-disciplinarity, relevance, rigour, and representation amongst Southern scholars. It discusses the SIRCA programme, its inception, opportunities, issues and concerns; the management process and evaluation. The section primarily underscores the need for ICTD research studies to be more analytical and empirical as opposed to descriptive and anecdotal. It also expresses the need for these studies to have more rigour and sophistication in ICT4D research methodology. We learn of various activities, including the mentorship collaboration, seminars, workshops, and research fieldwork and publication support that the SIRCA programme managed. The narratives on the management level highlighted the relationship of the mentors and PIs, the lessons learnt, and how the programme inspired the PIs to engage in ICT research and practice. The evaluation of the programme shares the programme's success in achieving its objectives.

The second section addresses the field-building reflections that emerged out of the interactions fostered by the SIRCA programme. In so doing, this section reflects the range of academic traditions represented in the programme through the mentors' lens. The section highlights the importance of ICT praxis and the links between research, instruction and professional development. However, research is not solely a straight pathway to theorising. The next chapter goes inside a "clean room" of research methods — structured, organised, systematic — but acknowledges the inevitable messy methods of data-gathering that transpire within the complexities of fieldwork. This is not to taint the integrity of each method but it is the reality of how several factors affect the context of the research being conducted, social or theoretical factors, at that.

The next chapter provides a comprehensive overview of ethics in ICTD research, particularly when dealing with the cultural factors in the communities where the work is embedded. The last chapter in the section champions the importance of an ICTD curriculum, drawing upon existing models from varied countries. An ICTD curriculum promises not only to train future researchers and academics but will also sustain the practice itself. Such a curriculum would produce a critical mass of practitioners who could contribute to various development, educational, government, and research organisations seeking knowledgeable and well-equipped professionals.

The third and last section is composed of research papers produced by the programme. With studies conducted in Cambodia, India, the Philippines, and Vietnam, the exemplars provide a concrete picture of the processes the researchers have undergone. SIRCA has served as a platform for the cross-fertilisation of ideas between first generation ICTD scholars and the new generation, between the experienced school and the emerging crop. This collaboration has occurred across academic traditions, cultural backgrounds and ethnic boundaries. SIRCA has led to a network of scholars that are bound not by schools of thought, ethnicities, age or gender, but by a sense of purpose to use ICTs in the service of the Millennium Development Goals. In this sense, participation in the SIRCA programme has brought higher order benefits to both its principal investigators, mentors and staff. Based on the studies, narratives and recommendations provided by this volume, we can see the extent of engagement to which one is required in pursuit of ICTD research. We conclude with a thought to the future, as SIRCA embarks on expanding the model developed in Asia to benefit researchers and communities in the Global South, particularly in Africa and Latin America.

References

Abraham, Reuben. "Mobile phones and economic development: evidence from the fishing industry in India." *Information Technologies and International Development* 4 (2007): 5–17.

Chib, Arul. "The Aceh Besar midwives with mobile phones project: Design and evaluation perspectives using the information and communication technologies for healthcare development model." *Journal of Computer-Mediated Communication* 15 (2010): 500–25.

Heeks, Richard. "Introduction: Theorising ICT4DResearch." *Information Technologies and International Development* 3 (2007): 1–4.

Jensen, Robert. "The Digital Provide: IT, Market Performance and Welfare in the South Indian Fisheries Sector." *Quarterly Journal of Economics* 122 (2007): 879–924.

Richardson, Don, Ricardo, Ramirez, and Moinul Haq. "Grameen Telecom's village phone programme in rural Bangladesh: A multimedia case study final report." In *Canadian International Development Agency*, 2000 <http://www.telecommons.com/villagephone/finalreport.pdf> (accessed 10 February 2009).

Subramanian, T. S. "Their own warning systems." *Frontline* 22 (2005): 15–28.

United Nations. *The Millennium Development Goals Report 2005*. New York: United Nations, 2005.

2

SIRCA
An Opportunity to Build and Improve the Field of ICT4D

Chaitali Sinha, Laurent Elder and Matthew Smith

INTRODUCTION

The spread of information and communication technologies (ICTs) across low and middle income countries (LMICs) in Asia has led to differential patterns of appropriation and use (Samarajiva and Zainudeen 2008; Butt et al. 2008). Some amongst these are expected, others less so; some contribute to positive social change, whilst others bear less favourable outcomes. These variances are embedded within, and influenced by, the interaction of social, political and economic systems that characterise different communities and livelihoods. The field of ICT for Development (ICT4D) is premised on the explorations, interpretations and analysis of these multi-dimensional and complex relationships. The resulting research findings and their communication has the potential to enhance equity and development outcomes, provided they are rigorous, relevant and useful to inform policy and practice.

However, some ICT4D scholars argue that the field has not lived up to its potential. Some claim the research is too descriptive at the expense of analysis (Heeks 2007); others emphasise the disproportionately large role of ICT4D researchers external to the countries and contexts they study (Coward 2007);

and many more bemoan the lack of theoretical underpinnings and the trap of disciplinary silos (Best 2010; Alampay 2009). A common thread emerging from this collection of observations and critiques from ICT4D research scholars relates to notions of academic rigour, relevance, and utility of the findings. The purpose of this short chapter is not to engage deeply in these debates, but rather to explain the emergence of one approach to address these critiques through strategies to fostering quality research by researchers based in the LMICs being studied. Critically, underlying this approach is the belief that ICT4D research should ultimately attend to relevant development priorities and equity concerns and subsequently be conducted and communicated in a manner that enables its translation from research to practice.

The challenges discussed inform the rationale for the research projects supported by the ICT4D programme area of Canada's International Development Research Centre (IDRC). For over 15 years, IDRC has responded to development priorities, as articulated by researchers based in LMICs, which explore both the positive and negative effects of ICTs in society.

The Strengthening ICTD Research Capacity in Asia (SIRCA) programme represents an IDRC-supported effort to address several of the common challenges confronting the field of ICT4D by developing a comprehensive approach to building the capacity of Asian researchers to enable them to produce quality research; i.e. research that produces findings that are rigorous, relevant, and useful. The following section briefly describes the genesis of SIRCA and how it fits within IDRC's history of supporting ICT4D research done by researchers in LMICs. Next is a discussion about the challenges that were identified by members of the ICT4D research community with regard to designing, implementing and communicating quality research findings, which segues into the presentation of an integrated framework of research capacity building that formed the basis for SIRCA. Finally, we conclude with a commentary on the centrality of translating research to practice for both SIRCA and ICT4D research as a whole.

IDRC'S INTEREST IN STRENGTHENING ICT4D RESEARCH

What do ICTs bring to the lives and livelihoods of the poor? Are ICTs levers or instruments to bring about a desired change? Do they act as a disruptive force in the change process? What social and cultural factors are critical for driving an ICT-enabled change process? These questions are just a sample of the myriad issues that need to be addressed if ICTs are to have a positive impact on people's lives. Research on them uncovers a range of responses depending on which communities are being studied, who is studying them, and what epistemological and ideological perspective they are coming from. Despite the difference of strategies, approaches and outcomes, IDRC has long maintained the need to support researchers from LMICs to design, implement and communicate these types of research studies and the resultant outcomes.

Much of ICT4D research suffers from work that lacks relevance and usefulness for the development purposes they are designed to inform and influence, for a variety of reasons. In the article titled *Theorising ICT4D Research*, Richard Heeks posits:

> There has been a bias to action, not a bias to knowledge. We are changing the world without interpreting or understanding it. Most of the ICT4D research being produced is therefore descriptive not analytical. It might make some interesting points but it lacks sufficient rigour to make its findings credible and it can often be repetitive of earlier work. It has a close-to-zero shelf life. The pictorial analogy of such work is that of stones being thrown into a pond, each one making a ripple but then sinking without trace (2007, p. 1).

In addition, there are some concerns about the reliance on anecdotal cases, success stories, disconnect from relevant theories, and limited attention to development outcomes that are influenced by ICT4D policies or practice (Chong 2011; Heeks 2007; Alampay 2009; Best 2010). Furthermore, as pointed out by Best (2010), whilst many ICT4D projects fail to produce outcomes, the problem lies not with the projects, but rather in the failure of the field to learn from the failures as well as from the occasional successes. Finally, but certainly not least, there is limited relative contribution from researchers in developing countries (Coward 2007), which can influence the relevance of research. For example, this can lead to 'extractive' research on development — rather than research for development. This is where developed country scholars engage in research on development — parachuting in and extracting data, only to leave the developing country and publish the findings. Even more egregiously, these findings might be published in a "closed" journal whose access cost makes it effectively inaccessible to those in the country where the research was conducted.

PLANNING ACTIVITIES INFORMING THE DESIGN OF SIRCA

For over a decade, IDRC has actively supported projects that strengthen the field of ICT4D research conducted in developing countries in an informed and systematic way. The planning activities leading up to SIRCA started in the latter part of 2006 with an initial roundtable of experts. This meeting was where the initial seeds were sewn for a systems approach to strengthening the quality and influence of ICT4D research studies done by researchers based in Asia. This was followed by two workshops that were twinned with regional or global ICT4D conferences. The first workshop took place in Manila in April 2007 immediately following the *Living the Information Society Conference*, whilst the second took place in Bangalore in December 2007 in conjunction with the *ICTD2007 Conference*. These meetings provided much of the creative thinking that shaped the initial ideas and strategies for SIRCA.

Workshop in Manila — April 2007

In preparation for the three-day workshop in Manila, a study was commissioned to gain a better understanding of the ICT4D research landscape in Asia. The evidence presented from the study titled "Research Capacity in Asia: A Literature Review on the Information Society" (Coward 2007) provided insights into the nature of scholarly publishing in the field, disciplinary affiliations of authors, as well as their organisational and geographic affiliations.

ICT4D research should by its very nature be inter-disciplinary, meaning that the field connects multiple distinct academic disciplines and/or schools of thought together in an effort to understand an emerging realm of enquiry. The growing prevalence and societal implications of mobile phones, PCs and the Internet have led to the interrogation of ICT issues from different sectoral perspectives and a wide range of academic disciplines. The distribution of publications according to departmental affiliation, shown in Figure 2.1, illustrates the cross-disciplinary interest in the field of ICT4D. These data show that nearly every social science and professional discipline publishes about ICT4D, with a stronger inclination in the applied sciences and to a lesser extent, social sciences. The findings should be interpreted with the caveat that departmental labels and classifications are not universal across all tertiary institutions.

The occasionally divisive disciplinary discussion within ICT4D is examined by Richard Heeks through a simple, yet effective, atomisation of the acronym 'ICT4D'. Specifically, he considers the 'I' as related to 'library and information sciences', the 'C' drawing from 'communication studies', and the 'T' in association with 'information systems' (Heeks 2007). Note that in Figure 2.1, there is a low level of author affiliations in the field of development — or the 'D'.

FIGURE 2.1
Author Department Affiliations (n=559)

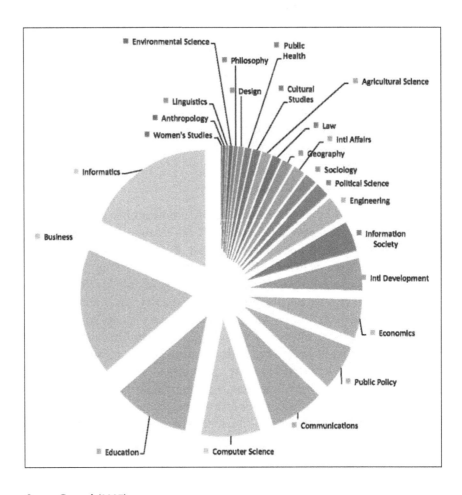

Source: Coward (2007).

This is not surprising to Heeks as he argues that little work has been done to date on ICT4D that links the 'D' to 'development studies'. Of course, interdisciplinary research has its limits as it is not feasible — nor would it likely be useful — to draw from every possible discipline that could apply to a particular study. However, its opposite can also be harmful as "discipline siloing restricts the creative thinking and diverse ideas that come from combinations across disciplines" (Best 2010, p. 50).

Other results from Coward's study examining different scholarly publications on ICT4D research in Asia showed that approximately two-thirds of the authors were based at organisations in regions outside of Asia (32% in North America; 25% in Europe; and 5% in Australia or New Zealand). Amongst the authors based at organisations in Asia, India (40%), Malaysia (16%) and China (14%) had the strongest representation.

A pivotal activity at the workshop held in Manila was a discussion of some of the core problems or challenges that are found in the field of ICT4D. After clustering the ideas that were shared by the group, three key challenges were portrayed using 'problem trees'. These structures utilise three main elements — notably the trunk(s): the core problem; root(s): the causes; and branches: the effect(s).

The three problem trees presented distinct but highly interrelated issues identified by the participants. Shown in Figures 2.2 through 2.4, the problem trees addressed the (1) Lack of rigour; (2) Challenges related to interdisciplinary research; and (3) Challenges related to collaboration. The ordering of the trees is not in any way representative of their relative weight or significance with regard to ICT4D research. Moreover, although the problem trees portray a certain degree of linearity, these should not be interpreted as rigid causal relationships; instead, their purpose is to represent the range of factors that influence and are influenced by the core problem.

Problem Tree 1: Lack of Rigour

Limited and varying levels of rigour found in ICT4D research was a recurring challenge discussed by the participants. Although much is written about the lack of rigour in ICT4D research (Heeks 2007; Burrell and Toyama 2009; Best 2010), there is less written about how rigour is defined in the field. Of course, this is not a question with a unique response. That being said, we believe the notion of rigour is closely related to issues of validity, reliability, trustworthiness and quality (Golafshani 2003; Davies and Dodd 2002).

The issue of limited rigour in ICT4D research, as presented in the problem tree, is illustrated in Figure 2.2. According to this depiction, the root causes that contribute to a lack of rigour include issues of data collection, capacity of researchers and the politics of using research findings. These lead to intermediary issues such as the ways in which data are analysed, the frameworks in which they are examined, and the level of support received to carry out certain studies. Stemming from these are the resulting causes for the lack of rigour found in ICT4D research studies; notably the paucity of reliable data, appropriate research models, limited institutional commitment

FIGURE 2.2
Lack of Rigour Problem Tree

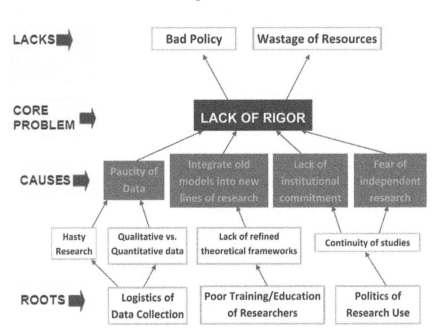

Source: IDRC-supported ICT4D Methodologies Workshop Report, Manila, 2007.

to carry out research, and a fear of pursuing independent research that may challenge dominant political forces. The problem stated as a 'lack of rigour' was then found to have negative effects on policy decisions and allocation of resources.

Problem Tree 2: Supporting Interdisciplinary Research

The second of three problem trees focused on the lack of interdisciplinary research in ICT4D. The root causes discussed in the formulation of this tree include matters related to the political, structural and institutional challenges defining the research environment; counterproductive variances in terminologies and assumptions used within and across different disciplines; as well as assorted research frameworks and methods that are utilised by researchers from a range of disciplines working in ICT4D. These root causes lead to challenges in aligning with shifting political agendas; the trap of disciplinary provinciality at the expense of interdisciplinary richness; and understanding how the frameworks and methods influence and shape priorities

FIGURE 2.3
Interdisciplinary Research Problem Tree

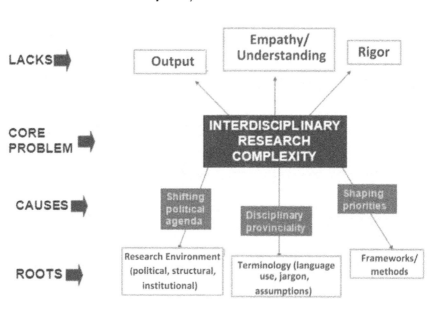

Source: IDRC-supported ICT4D Methodologies Workshop Report, Manila, 2007.

within a particular study. This leads to the 'trunk' of the problem tree, which is the challenge and complexities related to interdisciplinary research. Effects of this problem can be found in the quality of the outputs; the resonance of the findings (relating to empathy and understanding of the intended audiences and users); as well as the level of rigour and research quality.

Problem Tree 3: Challenges in Collaborating

The third of the three problem trees discussed at the workshop in Manila examined the challenges arising from a lack of collaboration in ICT4D research. Whilst collaborative research can have negative implications on the research process (Landry and Amara 1998; Cummings and Kiesler 2005), many argue that research collaboration is inherently 'a good thing' and should be encouraged (Katz and Martin 1997). Good inter-disciplinary collaboration, for example, often requires collaboration across disciplines.

This discussion in Manila on the lack of collaboration in ICT4D research took into account the different types of ICT4D researchers and different users of this research (see Figure 2.4). In particular, differences across researchers

FIGURE 2.4
Lack of Collaboration Problem Tree

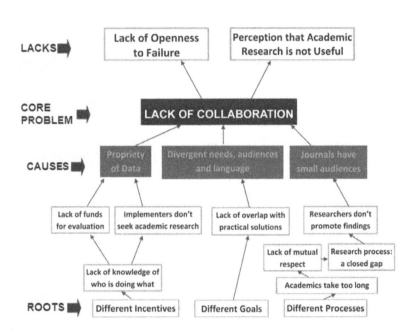

Source: IDRC-supported ICT4D Methodologies Workshop Report, Manila, 2007.

from traditional academic settings with those from contexts more closely linked to practitioners were discussed as a contributing factor to collaborative research efforts.

According to the tree, the root causes leading to challenges in collaborating include differing incentives, goals and processes. These can then lead to a limited understanding of other work in the ICT4D field, an inappropriate balance of academic research and practicable solutions, and a lack of mutual respect for the different types of ICT4D research that is underway. As a result, limited emphasis is placed on knowledge translation and using accessible language to reach the pertinent users and audiences for the research.

EMERGENCE OF INTEGRATED FRAMEWORK

The three problem trees related to a ((1) Lack of rigour; (2) Challenges related to interdisciplinary research; and (3) Challenges related to collaboration, represent distinct but highly interdependent and interconnected challenges.

Underlying each of them are assumptions of what constitutes quality ICT4D. The interdependencies and interrelationships amongst the three problem trees indicated the need for an integrated approach to strengthening ICT4D research capacities in Asia in a mutually reinforcing way. Figure 2.5 illustrates an interconnected framework of areas that were identified during the workshop as areas of emphasis to help address the different challenges portrayed in the problem trees.

The framework consists of four main parts: (1) Research Funding; (2) Training NGOs and Practitioners; (3) Field Building; and (4) Dissemination. These would be woven together by strategically timed and focused workshops and conferences. All elements feed into a core that supports building a body of evidence by "southern researchers" in ICT4D research that is high quality:

FIGURE 2.5
Integrated Framework

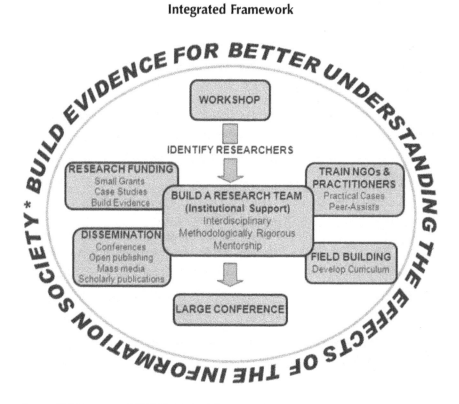

Source: IDRC-supported ICT4D Methodologies Workshop Report, Manila, 2007.

rigorous, relevant and useful for the development purposes that underpin the field of ICT4D.

The final piece of the puzzle, a strong mentorship component, came together at a follow up two-day workshop in Bangalore, India in December 2007. Leading up to the workshop was a call for research applicants. The selected applicants who were selected were invited to the workshop as well as to the *ICTD2007 Conference*. A number of established ICT4D researchers from Asia and beyond were invited to attend the workshop to provide input and guidance to the selected investigators. The richness and value brought about by the interactions amongst the researchers set the foundation to design a comprehensive and continued mentorship strategy within SIRCA. The role of the mentors were to collaborate with and guide the researchers on matters related to formulating informed research questions, developing an appropriate research design (drawing from different disciplines), discussing issues of ethics in conducting field research, analytical frameworks, as well as exploring the option of co-publishing the findings.

CONCLUSION

SIRCA was borne out of the deliberations discussed above with the desire to address the shortcomings of ICT4D research in a holistic manner. The SIRCA approach resulted in an ongoing learning process for the emerging researchers, through dedicated mentorship, peer-review processes, research capacity training, and opportunities to disseminate findings through different channels; such as conferences and peer-reviewed journals. We conclude with a short discussion on a key aspect of the SIRCA model that we believe has contributed to its success.

A common feature across all of the ICT4D research problem trees is the risks that can arise from poor use of the research; be it bad policy, lack of comprehensibility, or the perception of irrelevant research. In the beginning of the SIRCA process, however, the initial focus was mostly (although not entirely) on developing Asian researcher capacity to engage in rigorous, collaborative and interdisciplinary research. The underlying assumption here is that more useful research would emerge. As the project evolved, it would appear for the most part that this assumption has proven true; and consequently the project has evolved a greater focus on the importance of the translation from research to practice or policy (use).

In the world of research for development, it is easy to lose sight of the forest for the trees. Research can be technically rigorous, rich in theory, and elaborate in detail, but ultimately irrelevant to the intended change agents. This chapter is an appeal for ICT4D researchers to maintain focus on what we believe to be the key purpose of engaging in ICT4D research: use, the translation of research to practice.

References

Alampay, Erwin, ed. *Living the Information Society*. Singapore: Institute of Southeast Asian Studies, 2009.

Best, Michael L. "Understanding Our Knowledge Gaps: Or, Do We Have an ICT4D Field? And Do We Want One?" *Information Technology and International Development* 6 (2010): 49–52.

Burrell, Jenna and Kentaro Toyama. "What Constitutes Good ICTD Research?" *Information Technology and International Development* 5 (2009): 82–94.

Butt, Danny, Rajesh Sreenivasan, and Abishek Singh. "ICT4D in Asia Pacific: An overview of emerging issues." In *Digital Review of Asia Pacific 2008–2009*, edited by Felix Librero and Patricia B. Arinto. London: Sage, 2008.

Coward, Chris. "Research capacity in Asia: A literature review in the Information Society." Paper presented at Workshop on Research in the Information Society, 25–27 April 2007. Philippines.

Chong, Alberto. *Development Connections: unveiling the impact of new information technologies*. New York: Palgrave Macmillan, 2011.

Cummings, Jonathon N. and Sara Kiesler. "Collaborative research across disciplinary and organisational boundaries." *Social Studies of Science* 35 (2005): 703–22.

Davies, Deirdre and Jenny Dodd. "Qualitative research and the question of rigour." *Qualitative Health Research* 12 (2002): 279–89.

Golafshani, Nahid. "Understanding reliability and validity in qualitative research." *The Qualitative Report* 8 (2003): 597–607.

Heeks, Richard. "ICT4D 2.0: The Next Phase of Applying ICT for International Development." *Computer* 41 (2008): 26–33.

Katz, James Sylvan and Ben R. Martin. "What is Research Collaboration?" *Research Policy* 26 (1997): 1–18.

Landry, Rejean and Nabil Amara. "The impact of transaction costs on the institutional structuration of collaborative academic research." *Research Policy* 27 (1998): 901–13.

Patton, Michael Quinn. *Utilization-Focused Evaluation: The New Century Text.* 3rd ed. Thousand Oaks, CA: Sage, 1997.

Samarajiva, Rohan and Ayesha Zainudeen, eds. *ICT Infrastructure in Emerging Asia: Policy and Regulatory Roadblocks.* New Delhi & Ottawa: SAGE & IDRC, 2008.

3

MANAGING THE SIRCA PROGRAMME

Tahani Iqbal, Yvonne Lim and Arul Chib

The Singapore Internet Research Centre (SiRC) and the International Development Research Centre (IDRC) conceptualised the Strengthening ICTD Research Capacity in Asia (SIRCA) programme to improve the social science research skills of emerging Asian scholars in the information and communication technologies for development (ICTD) space. Given the perceived lack of rigour, scholarly representation from the Global South. and the need for inter-disciplinary research, the programme objectives intended to enhance research capacity amongst emerging scholars in Asia by supporting research that was scientific, replicable, generalisable, collaborative, and actionable (i.e. applied research).

Initiated in August 2008, the objectives of SIRCA (see Figure 3.1) were to: (1) Promote high-quality inter-disciplinary social science research in Internet development, e-services, new media use and social impact, and policy for the benefit and advancement of individuals, organisations, nation and society; (2) Support networks and linkages among researchers through a mentorship programme, as well as workshops and conferences to share knowledge and conduct training activities; and (3) Disseminate the research findings through such venues as academic journals, conferences and other relevant online and print media outlets. The programme was conceived by SiRC, based at the Wee Kim Wee School of Communication and Information (WKWSCI),

FIGURE 3.1
Snapshot of the SIRCA Programme

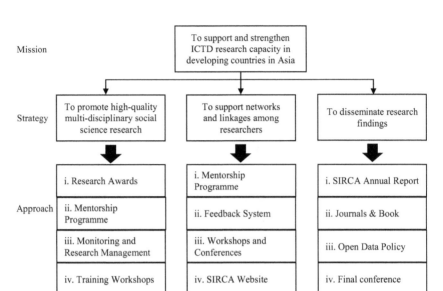

Source:

Nanyang Technological University (NTU), Singapore, and supported by the IDRC. SiRC was deemed a suitable location to incubate the SIRCA programme in, with its experience in conducting broad-based high-quality inter-disciplinary research related to new media and related technologies across Asia, and because of its wide-reaching networks for research collaboration, analysis, and technological development, with industrial, research and educational organisations in Singapore and overseas.

NEW APPROACHES TO CAPACITY BUILDING

SIRCA, a pioneer programme, was built on a three-pronged capacity building approach of providing financial support in terms of research grants, providing relevant research training workshops and exercises, and engaging grant recipients through a one-on-one mentoring arrangement. The need for mentoring surfaced at an IDRC workshop in 2007, where it was decided that a programme would be established as a test case for further capacity building projects. The SIRCA programme was amongst the first of its kind within the discipline to incorporate a mentorship arrangement in order to create opportunities for professional relationship building and mentoring between established researchers (mentors) and grant recipients (mentees) selected to the programme.

Unlike other capacity building initiatives undertaken by IDRC and similar development agencies across the world, SIRCA emphasised this unique feature in its programme and each selected grant recipient (referred to as the principal investigator or PI) was assigned a senior scholar whose role was to oversee and guide the progress of the research, with the potential for co-publishing the outputs. For the PI, this not only provided assistance and guidance to ensure the steady progress and completion of their project milestones but the opportunity to learn how to navigate the terrain of academia, or related fields, and move forward professionally. For the mentor, it was an opportunity to enhance his/her own personal and professional knowledge and networks whilst teaching and learning from the mentee.

MANAGEMENT

From 2008 to 2011, the SIRCA programme facilitated 15 research studies (12 grant recipients, and three graduate student awardees) of emerging ICTD researchers from eight Asian countries. These research projects covered a variety of topics addressing key developmental goals in agriculture, education, health, migration, livelihoods, and disaster-preparedness; for the benefit and advancement of individuals, organisations, nations and societies in Asia.

The SIRCA grant call (i.e. call for proposals) was launched in late August 2008. The call was disseminated via email to numerous researchers in various Asian universities, research centres, non-governmental organisations, associations and other groups relating to the field of ICTD. In addition, a website was set up for the call with the necessary information such as deadlines, submission instructions and related documents. In order to gauge the response in terms of the research topic as well as geographic location, a call for interest was issued a month prior to the actual call for proposals, requesting applicants to indicate their names and project titles. The advance notification helped in the management of the review and mentor-selection process by allowing the topics to guide their selection, rather than vice-versa. Secondly, the notification allowed feedback to those candidates whose topics were not compatible with or lacked relevance to the SIRCA objectives. For example, projects focusing on technological hardware development were advised to include a sociological perspective or were guided to other, more suitable, grant calls. 111 registrations of interest were received, which eventually translated into a total of 57 full proposal submissions by the deadline.

The final selection of all mentors and primary investigators (PIs) was made by the grants review committee (GRC), which comprised eminent academics, practitioners and IDRC representatives, from six countries (see Table 3.1). The review of proposals was conducted in line with international best practices of the double-blind process, wherein each proposal was assigned to two anonymous reviewers, to ensure maximum level of objectivity in the assessments. The personal information of the applicant was removed from the proposals to ensure anonymity. The allocation of proposals was based on the fit between the research project and the expertise of each reviewer, gleaned from their research specialisations, interests, and curriculum vitae.

To ensure consistency of reviewing across the global group of reviewers, the SIRCA programme created a project assessment decision matrix. This assessment framework aimed to ensure a standardised review process, by assigning specific categories which reviewers used to grade the proposals; broadly summarised as relevance, influence and knowledge transfer, as well as feasibility. The first category evaluated the identification of a research gap, research outcomes, relevance to the beneficiary community, medium-to-long-term sustainability, and contributions of outcomes to beneficiary community welfare. The second category of influence and knowledge transfer looked for beneficiary community support and commitment, practical application and impact of the research, unique contribution of the knowledge, and plans for dissemination of research results. The final category of project feasibility evaluated whether the objectives were clearly stated, whether the chosen

TABLE 3.1
Grants Review Committee

Reviewer	Affiliated Institution
Arul Chib	Nanyang Technological University
Chaitali Sinha	International Development Research Centre
Erwin Alampay	University of Philippines
Heather E Hudson	University of San Francisco
John Maurice Traxler	University of Wolverhampton
Jonathan Donner	Microsoft Research India
Kathleen Flynn-Dapaah	International Development Research Centre
Rahul De'	Indian Institute of Management
Roger Harris	Roger Harris Associates

method was appropriate, whether the project economics provided value for money, whether the objectives were achievable, and whether risks would be managed effectively.

All 57 proposals were vetted and evaluated by the GRC during the review meeting, with feedback compiled for each, both in the categories for those selected for funding, as well those who were unable to make the cut. Bearing in mind that the goal of the programme was research capacity building, SIRCA was determined to give feedback even to the non-selected proposals in order to help the applicants improve on their research proposal writing skills.

Fifteen proposals (see Table 3.2) were finally accepted for the SIRCA Grant Award, conditional upon stipulated changes to their proposal and/or budgets, based on feedback from the GRC. Specifically, PIs were advised to revise their budgets according to the SIRCA budget guidelines. The grant awardees were given a month to revise their proposals based on the feedback of the GRC. Simultaneously, selected projects were assigned to pre-identified mentors, so that the collaborations could begin from the start of their projects. Our experience suggested that the mentor-PI matching was a critical juncture in the programme; with the resultant ease of collaboration within the experience and output, such as co-authored publications, dependent on the right fit between the individuals involved. This is discussed further in the chapter, "Mentorship Evaluation".

Mentors were senior scholars, identified and selected through SiRC's networks to supervise and guide PIs from the onset of their research process, ranging from theoretical and methodological issues, to analytical tools and research dissemination methods. In all, 10 mentors were identified from six countries (see Table 3.2) to aid specific projects. Mentor site visits were planned and encouraged as part of the programme. The visits were crucial to contextualise the project, and more importantly, to strengthen the relationship between the PI and mentor. Further, with scheduling and communication infrastructure challenges faced by some PIs and mentors, the physical visits provided the best opportunity for collaborative work.

Mentors ensured that projects had not only an applied practical context but were grounded in theory, a necessity for publication in the best peer-reviewed journals, and for contribution to the scientific community. The mentor reviewed and provided feedback on data collection tools, as well as on the fieldwork experience during their site visit. Fieldwork experience by mentors was particularly important for the SIRCA programme as s/he had the ability to resolve logistical and ethical issues that might have occurred in the course of the research, which may not have been identified from afar, due to the multi-country management and collaborations involved.

TABLE 3.2
PI-Mentor Matches and Research

PI	Mentor	Research topic
Balwant Singh Mehta, India	Rahul De' India	Nature and Quality of Employment in ICT Sector in India
Devesh Kishore India	Vibodh Parthasarathi India	Empowerment of Farming Community Through Use of ICTs: A Study in the Indian Himalayan Region
Kanliang Wang, China	Arul Chib Singapore	Healthcare Information Systems in China
M. Sirajul Islam Bangladesh	Shaikh Abdus Salam Bangladesh	Agricultural Market Information Services (AMIS) and Its Implications on the Rural Community in Bangladesh: Theory and Practice
Ma. Regina M. Hechanova the Philippines	Ang Peng Hwa Singapore	Lifeline Online: E-Counseling and the Well-Being of Overseas Filipino Workers and Their Families
Mary Grace P. Mirandilla the Philippines	Alexander Flor the Philippines	The Filipino Blogosphere: An Emerging Alternative Venue for Political Participation in the Philippines
Md. Mahfuz Ashraf Bangladesh	Roger Harris Hong Kong	An Impact Assessment Framework to Evaluate the Effectiveness of ICT-led Development Project at Micro (Community) Level in a Developing Country
MJR David Sri Lanka	Vibodh Parthasarathi India	Learning from Challenges: An Image-based Mapping Process to Study Influences of Community-based ICTD Initiatives in Sri Lanka
Chivoin Peou Cambodia	May Lwin Singapore	Internet for Learning and Knowledge Building: Students' Perceptions and Utilization in Cambodian Higher Education
Phal Des Cambodia	John Traxler United Kingdom	Awareness on Sexually Transmitted Diseases (STDs): Showcasing a Mobile Solution Designed for Young Cambodian People
Pham Huu Ty Vietnam	Richard Heeks United Kingdom	Upper River Erosion and Landslide Protection Planning: An Integration of Erosion and Landslide Hazard Mapping Using GIS and Remote Sensing and the Value Chain Approach
Shefali Oza Nepal	Arul Chib Singapore	Telemedicine for Community Health Workers: An Analysis of the Effects of Connectivity on Health Equity in Rural Nepal
TB Dinesh India	Rahul De' India	Communities, Technology, Participation

In SIRCA's efforts to build research capacity amongst emerging researchers in Asia, the mentorship programme was an invaluable tool in consolidating relationships, both between PIs and amongst mentors, as well as between the mentors themselves. Indeed, this volume owes its genesis to the latter; borne as a result of serendipitous conversations amongst mentors, related to preserving the SIRCA experience for future generations of ICTD researchers, as well as those interested in capacity building exercises. In serving as SIRCA mentors, it was hoped that senior scholars would gain from guiding the research, development, and execution of the project, despite being located, physically, at a distance. Moreover, the bi-directional learning and exchange between mentors and PIs facilitated the outcome of stronger and more rigorous research findings that possibly led to increased publishing opportunities for both.

However, mentorship, as developed within the SIRCA programme, as a remote support tool, cannot be categorically defined as a success. Whilst some mentor-PI matches did not expand into greater working relationships, most did lead to fruitful exchanges and benefitted both partners immensely. Evidence of this is provided later. "By offering a blend of national and international expertise to emerging scholars in Asian countries, SIRCA devised a powerful mechanism of mentorship to inculcate research capacity," said one mentor. It was, however, brought to SIRCA's attention that PIs should and could have some say in which mentor gets assigned to them and this will be considered in future.

Other stakeholders involved in providing capacity building support to the SIRCA programme included four trainers, recruited to conduct workshops at the early stages of research. The first training workshop was conducted in July 2009 at a two-day event, which was the first time mentors met with their PIs face-to-face. The workshop aimed to equip the PIs with a clearer understanding of the literature, theoretical and methodological issues with regards to their projects as well as provide them with a clearer understanding of how their project related, and contributed, to the broader ICTD discipline. Particular emphasis was placed on refinement of the research question posed, and how this could add value to the existing body of knowledge in the ICTD domain. In addition, SIRCA hoped to provide PIs with the necessary information to adhere to the SIRCA Secretariat guidelines and highlight the practical issues and challenges involved in fieldwork.

The projected timeline provided for the mentor site-visit to occur next, so as to provide context and fieldwork experience to both the mentors and PIs. Following this meeting, mentors and PIs were expected to communicate regularly about the progress of the research. The two groups were brought together again about a year later. SIRCA's event, the *International*

Communication Association 2010 Mobiles Pre-conference: Innovations in Mobile Use in June 2010, allowed mentors and PIs to share progress on their research. As most projects had conducted their fieldwork, and were immersed in data-analysis and preparation for publication, a one-day SIRCA Workshop II, held the day before the pre-conference, focused on PIs' presentation skills and their dissemination plans, amongst other deliverables.

During the course of the research, PIs and mentors were given access to online libraries for use in their literature review, and were supported in disseminating their research findings at various international conferences and media outlets. Although a dissemination plan was not devised during the early stages of the programme, it was deemed important as the research work came to a close. This importance of policy and public influence via media dissemination was one amongst the few processes that emerged as a result of the inter-mingling of SIRCA stakeholders. For programme managers, it was important to allow for some flexibility to address emergent processes and thinking that occurred due to the interactions between various bodies of scholars, practitioners, researchers, and policy-makers during the course of the programme.

The SIRCA programme conducted objective evaluations, so as to learn from the experience, and to plan for improvement, both as course-correction to the existing programme, as well as for future initiatives. Two evaluations were conducted, one on the general programme and the other on the mentorship arrangement. Both evaluations provided insight into how the programme could be improved during and as well as in another reincarnation of the programme.

As part of SIRCA's mandate to build ICTD research capacity amongst emerging Asian scholars, graduate student awards (see Table 3.3) were offered to current or incoming graduate students of the WKWSCI at NTU. The SIRCA Graduate Award (Research) was designed to enhance opportunities for graduate students to conduct research. In return, the award required the recipient to support SIRCA-related work, in the hope that it would influence and benefit them in their research endeavours. The award assessment process required all applicants to first pass NTU's graduate admissions criteria. Thereafter, qualifying applications were submitted to IDRC for review and decision to award.

The SIRCA secretariat at SiRC was kept small, with a five-member team which comprised the Director, Assistant Director, Senior Manager, and two Programme Officers for research and finance respectively. The secretariat liaised with mentors and PIs and other partners throughout the programme, ensuring that any programme-related problems were dealt with in a professional and

TABLE 3.3
Graduate Student Projects

PI	Degree	Research topic
Komathi Ale, Singapore	MA	Evaluating the Impact of the OLPC Laptops on the Psychological Empowerment of Primary School Children in Rural India
Rajiv George Aricat, India	PhD	Acculturation of Migrants in Work Sector: Identifying the Facilitative Role of Mobile Phones
Thanomwong Poorisat, Thailand	PhD	Mobile Phones for Healthcare in Rural Thailand

timely manner. The secretariat also maintained an open knowledge policy throughout the project period enabling any mentor, mentee or partner to access the programme's internal working documents at any time.

OUTCOMES AND FEEDBACK

As a capacity building programme, the main objective of the SIRCA programme project was to develop the research skills of emerging Asian-based ICTD scholars. Based on the research work and its output, it is evident that the programme has improved the skill sets of PIs involved. Six of the 15 research projects were published in a special peer-reviewed journal issue of *Media Asia*, published by the Asian Media Information and Communication Centre (AMIC) in April 2011. In order to showcase findings of the research conducted by its PIs, SIRCA conducted a final dissemination conference in Thailand in April 2011. The audience comprised several international academics and practitioners, and the research was well received by all. Professor James Katz, Professor II of Communication, and Director, Center for Mobile Communication Studies at Rutgers, the State University of New Jersey had this to say,

> Thank you for hosting the most interesting and valuable event in Phuket last weekend. I learned a great deal from the presentations and my interactions with attending scholars. It was impressive to see how much has been accomplished, and how your goals are being developed to address the next stage of needed assessment and programmatic improvements.

It is clear that SIRCA has ultimately contributed to the growth of ICTD research in the process of training and developing the research skills of PIs. In addition to meeting the objectives of dissemination of PIs' research outputs through academic journals, conferences and online media outlets, SIRCA fuelled interest and motivation amongst its PIs to pursue further research and contribute positive outcomes to the research beneficiaries.

For instance, researcher Pham Huu Ty from Vietnam opened up a future research trajectory as he bid for a grant to open a centre focusing on climate change research which would have a positive impact on the research beneficiaries in Vietnam. Balwant Singh Mehta from India, whose research project studied the shared public access of telecentres, was one of four Fellows selected from Africa, Asia and Europe, to be part of the Amy Mahan Research Fellowship Programme to assess the impact of public access to ICTs (2010–11).

Additional outcomes of the SIRCA programme were to have the PIs' research incorporated into practice and for collaborative networks to blossom out of the mentor-PI relationship. This unique approach indeed made a difference to the quality of research produced by SIRCA PIs, and also resulted in progressive network-building between mentors and PIs. To quote Dr Roger Harris,

> My activities with SIRCA dovetail nicely with the other work that I do with universities, especially in bringing academics, practitioners and beneficiaries together so that each can better understand the others' perspectives as they work towards a common set of goals. My mentoring opportunity in Bangladesh is heading in this direction and I am now advising the University of Dhaka on further activities relating to this topic.

More statements from mentors describing how the SIRCA programme benefitted their work and helped to further their professional research expertise, with the following statements from Professor Richard Heeks, Professor of Development Informatics in the Institute for Development Policy and Management, the University of Manchester, and Associate Professor May Lwin, Associate Chair (Undergraduate Studies), NTU, respectively,

> My main work this year has been focused on a research topic triggered by the SIRCA mentoring process. On my field visit to Vietnam, I saw how climate change was increasing the problems of landslides and erosion, but also increasing the need and opportunity for ICTs to play a role. This led to a proposal to IDRC which I'm now working on:

a two-year project on ICTs, climate change and development. We are just in process of publishing a set of thematic papers reviewing the key areas in which ICTs can have an impact: adaptation, mitigation, monitoring and governance.

The SIRCA experience has helped to widen my regional network and collaborations in health communication especially in Cambodia where the mentorship was based. Thanks to SIRCA, I was able to include Cambodia in a recent study on an international health crisis communication project.

The programme also gave the mentors a refresher course on new research methods, as described by Professor Ang Peng Hwa, Acting Head, Division of Journalism and Publishing, WKWSCI,NTU, and Professor Alexander Flor, the University of the Philippines — Open University, respectively,

I was inspired by what I saw in [my PI's] class when they went over the transcript of the online counselling. She had combined her research project with a class seamlessly and meaningfully. Students were learning how to be better counsellors through 'live' cases that happened only a few days before. That "let's-do-it" spirit that was in the air inspired me to run a new course on Internet governance, which is my area of research. It is a lot of work as the whole area is new, without a proper [body of literature] but it's been refreshing and rewarding.

The SIRCA programme has served as a platform for the cross-fertilisation of ideas between first generation ICTD scholars and the new generation, between the old school and the new crop. And this collaboration has occurred across academic traditions, cultural backgrounds and ethnic boundaries. In this sense, participation in the programme has brought higher order benefits to both its principal investigators, mentors and staff.

In addition to achieving the goals set out by the SIRCA programme — that of developing research capacity to internationally recognised levels by engaging senior and junior scholars in a collaborative model, many PIs managed to outdo expectations. Three SIRCA PIs were awarded scholarships to pursue their PhDs, and others were honoured with awards and recognised for their SIRCA research work (see Table 4.4). PI Sirajul Islam from Bangladesh, says "the fellowship of SIRCA has not only given me an opportunity to help in conducting field work for my PhD study, but has also contributed to broadening my research capacity and network."

TABLE 3.4
SIRCA Capacity Building in Asia

Scholarships	Recognition of work	Research paper awards
Komathi Ale, PhD scholarship to USC Annenberg in USA in August 2011	Regina Hechanova, recipient of The Outstanding Women in Nation's Service (2010), Philippines, and the Cardinal Sin Book Award, Philippines	Rajiv Aricat George, top-three paper award at the Inaugural Honors Symposium for Asian Students in Communication Research held in City University of Hong Kong in November 2010
Chivoin Peou, PhD scholarship to University of Melbourne in Australia in July 2010	Mahfuz Ashraf, UNESCO Chair in ICTD scholarship to attend the ICTD2010 IEEE/ACM Conference in December 2010	Phal Des, Distinguished Paper Award at IEEJ Japan-Cambodia Joint Symposium on Information Systems and Communication Technology 2011 (JCAICT 2011) in January 2011, in Phnom Penh, Cambodia
Pham Huu Ty, Nuffic Scholarship for PhD at Utrecht University, Netherlands in May 2010	Sirajul Islam, Reviewer, Chair and Moderator at ICTD2010 IEEE/ACM Conference in December 2010; Reviewer at the 44th Hawaii International Conference on Systems Science (HICSS) in January 2011, at the 2nd International Conference on Mobile for Development, in Kampala, Uganda in November 2010, and for the Journal of Information Technology for Development (ITD)	

Chivoin Peou from Cambodia was awarded Australia's Endeavour Award for PhD study at the University of Melbourne with research emphasis on Internet and youth issues. Building on the experience of working on his SIRCA project understanding the internet usage of Cambodian youth, and developing an interest in ICTD, Peou successfully obtained the Australian scholarship to further his knowledge and investigation in the topic of ICTD. To quote Peou, upon receipt of the PhD scholarship:

> The SIRCA grant I received between 2009 and 2010 not only offered me a chance to research an under-researched yet significant area of ICTs in Cambodia, but also paved my way into a committed research and academic career. The SIRCA experience has both aided my admission into the PhD programme at the University of Melbourne and inspired my research and academic vision within my current PhD programme.

Further, Peou's project has been documented as an active research project for the Department of Media and Communication at the Royal University of Phnom Penh, which plans to set up a media research programme that builds further on the project experience. These outcomes attest to the capability of the SIRCA grant programme to nurture and sustain research capacity in ICTD.

PAVING THE WAY FOR THE FUTURE

Based on the feedback and the response received from all stakeholders involved, it is clear that the SIRCA programme has had a significant impact in the field of capacity building. As such, it is hoped that the next round of capacity building efforts will address a larger pool of emerging scholars in the Global South. There is indeed a need to inculcate high quality research skills in order for scholars from these regions to produce top-notch quality academic work. This will help to reduce the lack of representation of Global South researchers in the ICTD field.

Based on the experiences and successes of the SIRCA programme, IDRC and SiRC have set out to take the programme to a much larger group of emerging scholars from Africa, Asia and Latin America. Collaborating with the Instituto de Estudios Peruanos (Institute of Peruvian Studies) in Peru, and the University of KwaZulu-Natal in Durban, South Africa, the programme was launched in the latter half of 2011. The second programme, aptly titled Strengthening Information Society Research Capacity Alliance (SIRCA II), will focus on research in the Information Society field.

Following in the footsteps of its predecessor, and learning from its successes and failures, as well as recommendations and feedback from participants — both mentors and mentees — SIRCA II will also follow a three-pronged approach to capacity building, with increased emphasis on regional collaboration and real world (policy) impacts. Geared with the knowledge from the first round, it is likely that that the programme will continue on its path of building research capacity the world over, and developing a global network of like-minded scholars. Hopefully, SIRCA will continue to achieve key objectives globally, similar to the experience of the first round in Asia, as stated by PI Mary Grace P. Mirandilla from the Philippines,

> SIRCA has provided not only a platform for research, but a venue for enriching exchanges, the much needed support for scientific and relevant inquiries, and a community of passionate individuals who are committed to improving lives with the aid of information and communication technology.

4

PRIMARY INVESTIGATOR AND MENTOR PERSPECTIVES OF SIRCA

Ann Mizumoto

EVALUATION

The Singapore Internet Research Centre (SiRC) committed in establishing SIRCA as one of the best ICTD research grant programmes in Asia, commissioned an external consultant to conduct two evaluations during its pioneer round. Given the ambitious objectives of the programme — to enhance research capacity in the region, to create a space for dialogue on ICTD social science research issues, to connect emerging and senior ICTD researchers through a mentorship scheme, and to raise the profile of ICTD work coming from Asia — SIRCA's management was ready to grasp the lessons learned and to implement relevant recommendations that the evaluations proposed. It was also an opportunity for the staff, entrenched in the day-to-day administration of the Programme, to get a big picture of their work.

The first evaluation was formative — it focused on improving the programme — and covered a period of two years and four months, from the programme's inception and conceptualization in March 2008 to July 2010 after SIRCA's second workshop, which brought together mentors and investigators from all countries and serendipitously provided an opportunity to collect data for the evaluation. The evaluator worked with SIRCA's management to select the areas of study, to articulate the uses of the evaluation, to identify the primary users of the evaluation, and to define the key evaluation questions. The Grant Review Process was one area for evaluation because SIRCA's management received feedback about the short review time of proposals which may have affected the application screening process. The Mentorship Programme was the second evaluation area as it was a unique component of SIRCA that went beyond a typical research grant programme. SIRCA embraced a vision of professional support and career formation for emerging ICTD researchers.

The Principal Investigator (PI) was linked to a senior researcher (Mentor) who would provide continuous guidance, support, and direction throughout the project. It was hoped that the Mentee benefited from this interaction with an experienced researcher while gaining confidence and skills to contribute to the field of ICTD through rigorous research, publications, dissemination, and collaboration with the Mentor. The third and final evaluation area was the SIRCA Workshops and Conferences which management could organise and immediately rectify with anticipated planning. The evaluation methodology relied on surveys that were emailed to all PIs, mentors, GRC members, trainers, and SIRCA staff. The respondents were given the option to return the survey via email, to answer it over the phone or in person. All interviews were recorded for accuracy. In addition to being an observer at SIRCA's Workshop II and Mobiles Preconference in June 2010, the evaluator spoke informally with mentors and PIs at these events. To facilitate the evaluation, the Secretariat shared internal documents with the evaluator such as application forms, progress reports, final reports, project amendments, quarterly and biannual feedback survey responses, email correspondences, and video recordings.

In light of the results from the first programme evaluation, it was deemed valuable to conduct an in-depth evaluation of the mentorship model as issues and questions for further clarification surfaced. The Secretariat was keen on understanding the Mentor-PI relationship so as to conceptualise a mentorship model that met the needs of emerging researchers. The second evaluation was entirely qualitative, relying on personal interviews conducted by the evaluator and the Secretariat staff with the use of a recorder and an interview schedule. PIs and mentors participated in individual interviews, and where possible, joint PI-mentor interviews. Most of the interviews were conducted at the project site but the evaluator also spoke to PIs informally during SIRCA's final dissemination conference in April 2011.

THE SIRCA PROGRAMME EVALUATION

The first key evaluation question had asked to what extent did the Grant Review Committee (GRC) select the most appropriate candidates to receive the SIRCA grants given the available time and resources. A closer look at the Grant Review Process found that SIRCA had faced some constraints as it was starting up the programme, leading to a delay in the call for proposals and subsequently, the submission period for applications. SiRC was running the programme under IDRC's umbrella but it also had to comply with guidelines stipulated by the Wee Kim Wee School of Information and Communication, which was in turn under the legal and administrative regulations of Nanyang Technological University.

Setting up the programme for the first time during a period of staff shortage pushed the proposal submission deadline back by several weeks and ultimately led to what GRC members felt was a short period of one week for each person to review about a dozen applications. In spite of time and resource constraints, SIRCA did a commendable job in outreach as it was able to advertise its programme on the websites of ICTD organisations, international ICTD conferences, governments, multilateral organisations, research institutes, think tanks, and universities. From the 111 online interest indication registrations, 57 full proposals were submitted from diverse countries and on various topics such as the internet, mobile phones, rural development, education, agriculture, political science, psychology, migration, disaster management, and healthcare.

The evaluation recommended that more advertising to non-academic entities doing development work can be done as SIRCA aims to understand and resolve development challenges using ICT. In addition to a longer review period and fewer proposals per GRC member, SIRCA can increase the rigour of the screening process by refining the application form and interviewing shortlisted candidates to assess their motivation, commitment, research experience, and language skills. The evaluation showed that PIs who started out with little research experience were the ones who had the most 'value added' from SIRCA as they learned how to administer a grant, to supervise staff, to manage time and pressure, with some going on to pursue further studies.

The second evaluation area was guided by the question of how SIRCA's mentorship programme facilitated learning and collaboration between emerging and established researchers. SIRCA assessed the geographic and thematic preferences of the mentors who applied to the programme and assigned the PIs to them accordingly. Matching was not perfect as some mentors were assigned to projects that were not completely within their expertise, but they compensated their unfamiliarity with the topic or region with good interpersonal mentoring and coaching skills. While communication was done on a needs-basis through email on most occasions, the mentor site visits were the highlight of the projects as mentors were able to better understand the on-the-ground challenges. Site visits improved working relations and were most valuable to mentors and PIs facing communication infrastructure and language issues. Being based in the same country, however, did not increase the number of meetings. Site visits were so essential that most PIs and mentors needed more than one visit, to which the Secretariat immediately responded and funded a second one.

The third and final evaluation area looked at SIRCA's capacity building activities — workshops and conferences. The question was whether these activities facilitated the publication and dissemination of research findings to the wider world. The evaluation found that not all PIs had the same level of research experience. Some were just finishing their Masters degree, while others were overseeing a team of researchers. This made it difficult for the Secretariat to create a workshop that satisfied everybody's needs. A general consensus among PIs and mentors was that SIRCA needed to hold at least two workshops and one conference during a programme cycle.

SIRCA's first workshop had four trainers who taught ICTD theory, quantitative data analysis, qualitative research methods, mixed methods, and research ethics. Break-out sessions were further arranged by topic (agriculture/ rural, education/political participation, socio-economic/psychological development, and healthcare). A second workshop and a Mobiles Preconference were organised immediately preceding the ICA Conference 2010 with funding for the first day of ICA so as to allow the PIs to gain international exposure to scholars doing ICTD research.

The participants appreciated the thoughtfulness of SIRCA to attach the workshop and preconference to such a large global academic conference. The second workshop invited mentors to a panel that focused on the final phases of the project, namely, publication and dissemination. Geographic and ICTD domain specificity, more depth to the topics, and more time to interact with the mentor at the workshop were suggested by the PIs. People from a variety of backgrounds i.e. journalists, development practitioners, and policy-makers, for example, can also impart hands-on skills at the workshops such as how to write opinion pieces and articles.

The SIRCA programme evaluation unveiled several issues that were subsequently re-examined in the mentorship model evaluation. First, the PIs expressed concern at being assigned to a mentor who was not very familiar with the research topic. Second, a few respondents felt that they needed more opportunities to communicate with other PIs and mentors, as the SIRCA workshops were their only medium of interaction. The second evaluation followed up with the utility of a social networking space to mitigate this sense of isolation. Third, the mentor honorarium issue was brought up and the mentorship evaluation probed respondents about incentives to participate in the programme. Finally, the idea of a publication and dissemination strategy for the PIs was revisited in the second evaluation. Some respondents in the first evaluation felt that SIRCA was not in a position or did not have the intent to address the research-to-practice gap. It was more important to get emerging researchers to publish in peer-reviewed journals and to get them recognised in the academic community before addressing policy and practice issues. In contrast, some respondents felt that SIRCA's social science research had an important role to play in resolving ICTD and development issues.

THE MENTORSHIP MODEL EVALUATION

Before revisiting these issues that emerged from the first SIRCA evaluation, the mentors themselves were asked to set the ground by defining the "ideal mentor". Unsurprisingly, the characteristics encompassed the roles and responsibilities of the mentor, including personality traits that would best facilitate the learning process for the mentee. The mentors articulated 'prerequisites' such as past experience supervising mentees, fieldwork experience, publication track, grant management experience, thematic and geographic expertise, ability to travel, familiarity with ICTD literature, good sense of timing for interventions, and good judgment about the use of resources given the PI's context.

The PIs gave a similar list but with more "people skills" such as approachable, supportive, motivational, inspiring, encouraging, patient, culturally-sensitive, proactive, and timely. The PIs felt that mentors had an influence on all stages of research — theory and conceptualisation, fieldwork and execution, and finally, publication and dissemination. Although they did not directly participate in fieldwork, mentors helped craft data collection tools and gave logistical tips for fieldwork. Mentors who were familiar with the publication landscape of specific ICTD domains such as mobile phone technology or e-health guided PIs to indexed journals of that domain and recommended relevant ICTD conferences to disseminate the findings.

Despite the seemingly clear understanding of a mentor's role, mentoring in practice was ambiguous. Both parties felt unclear about the extent to which the TOR and the MOU could be enforced. Mentors were unsure about their moral authority as a 'monitoring person' or 'police' versus a close and committed guide like a PhD advisor. Most PIs and mentors did not discuss their expectations of each other and of the project. They felt that a meeting organised by SIRCA early in the project would have been beneficial to clarify expectations. In addition to refining the duties of the mentor in the TOR and MOU, any potential conflict of interest issues can be clarified with the Secretariat at this early stage.

While the honorarium was mentioned as 'symbolic' in the first evaluation, the second evaluation confirmed that mentors participated in the programme for intellectual and academic pursuits. Monetary compensation was secondary to what mentors could gain from the programme — the opportunity to deepen their current research interests, to engage with cutting-edge research, and to be part of a global ICTD community. Higher honorariums may attract famous senior researchers who can increase SIRCA's profile but they may not be able to commit so much time for the PI. Matching the mentor to a project that fit closely to their expertise was therefore crucial on two counts — to maximise the PI's experience as a mentee and to increase the chances for co-authored publications as mentors would be more willing to write on a topic they are familiar with. More collaborative work between the PI and the mentor was anticipated in SIRCA but according to the respondents, it should continue to be a natural process instead of a mandatory programme output. Collaboration depends, in part, on the type of mentorship model that SIRCA adopts. Several model suggestions are illustrated in the second evaluation — for example, through assignment, PI nominations of mentors, a panel of experts, a PI-mentor-collaborator trio, or consultations with experts in parallel to an assigned mentor.

Communication is the key factor that underlies the SIRCA mentorship model, just as in any personal and professional relationship. In addition to communication between the mentor and the PI, the evaluation noted two other levels of communication — communication within the SIRCA cohort, and communication from the researchers to the wider world. In terms of communication amongst PIs and mentors, the first evaluation had recommended the use of a social networking site such as *Facebook* to alleviate the sense of isolation felt by a few respondents due to their remoteness. Instead of corroborating this perception, the second evaluation divulged that most PIs and mentors did not feel comfortable using a social networking platform, and in fact, did not feel the need to do so as long as they had each other's email addresses. They preferred communicating, or actually viewing each other, through an interactive platform like an online project management tool where everyone could track the projects on a timeline and the PIs could update their work or see how other researchers were doing.

At the second level of communication, both the first and second evaluations emphasised the importance of disseminating the research findings to academic and non-academic audiences. More was desired in terms of publication and dissemination to mass media that reached decision makers and the general public. SIRCA's research projects would have had a broader impact on policy and practice if it had been written in a less technical manner than for peer-reviewed journals. Thus, a discussion on a publication and dissemination strategy early in the programme was suggested where the mentors and other experts could elaborate on the various dissemination paths and talk about non-scientific styles of writing, perhaps providing practical help throughout the programme. The first SIRCA round resulted in more academic publications and disseminations from the PIs which the Secretariat, nevertheless, should continue documenting as successful outputs.

CONCLUSION

The evaluations of the SIRCA programme and the mentorship model served to raise awareness about the strengths and opportunities for improvement of the programme. The Secretariat is keen on maximising the utility of the evaluations and has already implemented several recommendations to modify processes and activities to fully achieve its capacity building mission. The diversity of people and ICTD themes brought under SIRCA has made the experience rich for all participants. There is no doubt that the growing 'SIRCA Family' will become an important player in the global ICTD community. The Secretariat's continued attention to monitoring and evaluation was key in identifying areas that need reform, thus allocating resources accordingly and establishing a grant programme committed to transparency, accountability, and excellence.

Key dates:
First SIRCA programme evaluation implemented November 2009 – November 2010
Mentorship model evaluation implemented December 2010 – May 2011

SECTION II

Research Perspectives:
Theoretical Reflections by Experts

5

ICTD PRAXIS
Bridging Theory and Practice

Alexander Flor

There is nothing more practical than a good theory.

Kurt Lewin (1952)

INTRODUCTION

This chapter elaborates on the theoretic dimension of my engagement with the Strengthening ICTD Research Capacity in Asia (SIRCA) Programme.

A distinctive feature of SIRCA is the appointment of mentors along with principal investigators in the research process. When enlisted into the programme as a mentor, amongst the many assumptions that I harboured was that research should lead to theory and that theory should result in further research. After all, should not a research capacity building programme help develop a robust theory building agenda? Should not the programme provide incentives to researchers to construct and validate theory?

In August 2009, I was asked to share my thoughts on these assumptions with the first batch of SIRCA grantees, now considered principal investigators (PIs) of information and communication technology for development (ICT4D or ICTD) phenomena. These assumptions, however, give rise to two issues.

AN ICTD THEORY BUILDING AGENDA?

The first issue deals with the nature of the field. ICTD has originated from, has been borne out of, and has been thriving in *practice*. For more than half a century, ICTs from the humble rural radio to the most sophisticated knowledge networks have been applied in the service of national development and, more recently, in support of the Millennium Development Goals (MDGs), generally without adopting solid theoretical frameworks or foundations. ICTD theory came *ex post facto*, more of an afterthought to help explain processes and outcomes, rather than *ex ante*, to be validated by research. It was hardly used as a basis for planning and practice. The dominance of applied research, mostly attendant to project monitoring and evaluation, attests to this. So does the preponderance of ICTD fugitive papers or grey literature produced by international development agencies over scientific articles published in technical journals.

The reasons for this trend are uncertain. It may have been a function of the immediacy of the societal problem being addressed (i.e. food security, poverty, pandemics, etc.), generating a sense of urgency that could not accommodate the luxury of theory building. The MDGs have a timeframe, which does not automatically freeze with the act of theorising. Perhaps, the rapid obsolescence of technology allotted a shelf-life to theoretical relevance as well. For a time, even the availability of published papers that have undergone rigorous double-blind peer-review have also been questioned, primarily since many of the ICTD publications have yet to be indexed in key scientific databases and hence were not considered reputable.

Alternatively, researchers may have found earlier theories, drawn from the available papers, usually inadequate as Rahul De' argues elsewhere in this book. ICTD inquiries were, more often than not, *transdisciplinary*, blurring the boundaries between information technology, information science, communication science, and development studies; thus the difficulty in establishing a theoretic anchor.

The nature of the discipline, if we may call it as such, has deposited theory in the backseat. ICTD presupposes the application of technology for a socio-economic purpose, hence adopting a purposive, practical rather than objective, theoretical slant. We may argue that communication science has yielded major theories applicable to ICTD. Yet, the products of these theory building endeavours have not been fully embraced by ICTD planners, policy makers and implementers. Can we then satisfactorily situate theory within the ICTD arena?

The second issue deals with the nature of the programme itself. SIRCA cannot be associated with any one specific school of thought or tradition.

A theoretical focus or track cannot be imposed upon the SIRCA research community building process. The variety and diversity of traditions under which ICTD research may be situated can make the link to theory unwieldy and tenuous. The SIRCA programme was meant as a platform for the cross-fertilisation of ideas between first generation ICTD researchers and the new generation, between the old school and the new crop. And this collaboration should occur across academic traditions. SIRCA would thus lead to a community of scholars that are bound not by an academic tradition but by a sense of purpose to use ICTs in achieving the Millennium Development Goals.

These two issues accounted for the initial hesitation in pushing the theory building agenda at the onset. Nevertheless, there were a number of developments on the ground that necessitated its *de facto* adoption.

MIDTERM MDG INDICTMENTS ON ICTD

When the international development assistance community marked the midpoint of the MDG timeframe in 2006, critical questions were raised on the impact of ICTD. The correlation between ICT infrastructure and Gini coefficients or poverty indices in Southeast Asian economies may have been proven earlier at the macro level (Flor 2001) but a causal link has not been established. Anecdotal evidence on the successes of ICTD at the project level likewise abound. However, three questions are yet to be answered satisfactorily: "Has ICT4D contributed significantly to poverty alleviation?" "What has been the overall impact of ICT4D in agriculture, in education, in health, in governance, in natural resources management?" "Where is the evidence?" The international development assistance community turned to research and, subsequently, to theory for answers.

ICTD began to be associated with the social promise of poverty alleviation at the *2000 Okinawa Summit of G8 nations.* Since then, however, many still remain unconvinced that it can bring about large-scale societal impact on poverty and the development process. At the time of SIRCA's conceptualisation, there was a marked hesitance amongst key players within regional financial institutions to invest in ICTD loan projects in Asia, particularly those that were non-infrastructure related. The return on investments of ICTD projects was not direct and considered uncertain.

The evidence presented by ICTD advocates was likewise contentious. In the *eIndia 2008 Conference*, a joint secretary in the Indian Ministry of Panchayati Raj likened survey evidence of telecentre contributions to economic development, to that of survey evidence of the number of tigers in India, (i.e. conclusions based on sightings of tigers' paw marks rather than the sightings of the species). In the same conference, a member of a keynote panel diplomatically stated, "The impact of ICT on development has not been very clear." The clamour for an evidence-based approach in ICTD was gaining headway and a SIRCA programme with a robust theory building agenda was needed to address it.

With this as an underlying premise, I shared my thoughts on the need for ICTD theory with the initial batch of PIs during the first SIRCA workshop in August 2009. At the back of their minds there were a number of questions on theory (both articulated and unasked) that I will endeavour to answer here.

QUESTIONS ON ICTD THEORY
Why should we theorise?

Why do research at all? Simply stated, we conduct research to add to the existing body of knowledge on a field of endeavour. A three dimensional body of knowledge is made up of theory, policy and practice: theory for the scientist, policy for the decision maker, and practice for the professional. Furthermore, it is implied that research coupled with reflection is the prerequisite for *praxis*, the marriage of theory and practice (Freire 1970).

From the perspective of practice, we do research to build up our procedural and methodological knowledge. From a policy perspective, we do research to determine the most appropriate course of action or guidelines to be adopted within certain parameters. However, from a purely scientific perspective, we conduct research to construct, validate and supplement theory. Hence, there is the primary preoccupation for rigour, objectivity and trustworthiness of data and procedures.

Why then do we theorise? We do so to understand, explain and/or predict phenomena unless we adhere to the postmodernist view that we cannot fully predict anything. Nevertheless, almost everyone, postmodernists included, agree that we need to theorise because we need to *know* on the basis of empirical evidence.

What do we currently know about ICTD? Is there an existing body of knowledge on ICTD?

Yes, there is, but this body of knowledge is skewed towards policy and practice. Much of what we currently know about ICTD has been generated by applied research. However, we can still mine, harvest or cull some sound theoretical constructs from these.

In 2008, I conducted a scoping study on ICT for rural livelihoods in Southeast Asia commissioned by the International Development Research Centre (IDRC) (Flor 2008). It employed literature reviews, the analysis of secondary data and key informant interviews. The findings just about summed up most of what was currently known on the subject and these I shared with the PIs.

Firstly, the study concluded that ICTD interventions in rural communities increased capacities of groups (particularly women) to offer marketable skills. Another known ICTD outcome is increased awareness and attainment of basic services by communities as well as increased linkages with government service providers. This supported an earlier conclusion made by International Telecommunications Union (ITU) researchers that ICT infrastructure in the developing world tended to extend from the backbone to the peripheries and on to the last mile through the participation of agencies involved in basic services such as agriculture, health and education (Calvano 2004), which supply both intermediaries and content.

Additionally, the study confirmed that ICTD interventions indirectly increased agricultural production and incomes amongst rural families. ICT related services also led to additional revenues to primary livelihoods such as retail outlets, local eateries and community transport. ICTs resulted in the development of support mechanisms and locally driven enterprises such as technical supply chains, marketing alternatives and other services. Furthermore ICTs increased social capital generated through knowledge sharing and networking amongst groups, particularly women.

Lastly, the scoping study found that ICTD interventions were likely to produce these outcomes on a sustainable basis when ICT livelihood projects: are run as secondary instead of primary livelihoods; are run by women groups; are linked to basic service providers; use intermediaries and personalised technologies; include a capacity building component; operate as public-private sector partnerships; and are neither technology-nor donor-driven.

Note that these findings not only help us in practising ICTD. They also contribute to our understanding of and allow us to explain ICTD phenomena. Nonetheless, much of these can only be called theoretical constructs or validated hypotheses instead of full blown theories. For the more sophisticated stuff, the PIs were given an overview of traditions that thus far guided ICTD research.

TRADITIONS OF FIRST GENERATION ICTD RESEARCH

Although ICTD, as a field, is borne out of practice, first generation ICTD research did not evolve independent of theory. Consider that ICTD is a product of the convergence of three areas of study with robust theoretical foundations: information technology, communication science and development studies. Early on, ICTD-related research and, subsequently, theory have been established along the lines of the development communication, and information economics disciplines. However, other traditions soon became engaged in the ICTD arena. Although not as exhaustive as may have been desired, I had a list of five traditions that I had personal knowledge of and involvements in, which I shared with the SIRCA PIs in August 2009.

Development Communication. In the 1970s, development communication theorists exemplified by Nora Quebral (1974) laid down the basic argument on how communication processes led to societal development. One of my earliest involvements in this area was a World Bank-funded project, Communication Technology for Rural Education (CTRE) led by Quebral herself and implemented between 1976 and 1981. This project was quite exceptional since it adopted a theoretical framework founded on communication science.

The development communication tradition was an offshoot of the agricultural information and communication tradition that prospered in American land grant universities, albeit applied within the context of developing societies. Hence, for a long time, development communication was positivist in approach and hypothetic-deductive in method. Theoretical frameworks were determined *ex ante* and validated empirically through empirical research. The dominant frameworks adopted then were Roger's diffusion of innovations, Lazarsfeld's two-step flow hypothesis, De Fleur's media dependency theory, and Katz's uses and gratifications theory.

Information Economics. Almost chronologically parallel with research in development communication was information economics. Also in the 1970s, economists such as Marc Porat (1978) and Meheroo Jussawalla (1989) argued that information had become the dominant commodity and the most critical resource in contemporary societies. It was Porat and his senior collaborator Fritz Machlup who introduced the Agricultural-Industrial-Information Age(s) trichotomy as well as the concept of the *information society.* Information society is defined as a society whose economy is information based, i.e. the majority of its workforce is made up of information workers or the greater part of its GNP may be attributed to information products, services and labour

(Porat 1978). Unfortunately, this last term is misused and abused these days — its usage in the *World Summit for the Information Society* or in conferences such as *Living the Information Society* contradicts the fact that the Philippines and much of Asia are agricultural societies in the Information Age.

My 1986 dissertation titled, *Two Faces of the Information Age in a Developing Country* was based on the Porat-Machlup framework. On the other hand, Jussawalla became a senior collaborator whilst serving as a research fellow at the East West Centre. Our research in 1989 focused on the informatisation of ASEAN economies. It may be said that information economics theorising branched out from the traditional labour economics school of thought. However, instead of the conventional agricultural-industrial-services labour sectors, Machlup and Porat proposed a slightly different classification: the agricultural-industrial-information sectors. Nevertheless, it adopted similar approaches, methods and analyses.

Knowledge Management. There is reason to believe that the current discourse on ICTD originated from a group of pre-World War II Austrian-born academics. This school of thought, known as knowledge economics, developed in the 1930s as a branch of Friedman's liberal economics. Its luminaries were Hayek (1936) and Schumpeter (1912, 1942). The tradition situated knowledge as a major economic variable but was overshadowed by the Keynesian school in the 1940s. It was resurrected in the 1970s by Machlup and Porat albeit under a new brand, information economics.

Then in the mid-1990s, Nonaka and Takeuchi (1995) published their seminal work on knowledge creating companies in Japan. This was followed by several papers on knowledge management by Davenport and Prusak (1998). In 1996, the World Bank established its Knowledge Management Programme under Stephen Denning. In Asia, research and development organisations such as SEAMEO-SEARCA followed suit. Soon, knowledge management for development (KM4D) programmes were being established by regional and international development agencies claiming that knowledge and ICTs can become the great equaliser in developing societies.

Critical Theory. However, it was largely because of the inequities attendant to the Information Age that the critical tradition found relevance in ICTD inquiry. Research questions on the Digital Divide, the information rich and the information poor, media imperialism and others were being answered from Habermas' power relations in communicative action (1983) and Schiller's communication and cultural domination (1976) frameworks. Coincidentally, I received the publisher's proof of my manuscript titled

"Developing Societies in the Information Age" prior to the August 2009 workshop. I took the liberty of having the proof photocopied for distribution to the PIs.

The above traditions manifested in the first 50 years of ICTD-related research. I concluded my discussion with the SIRCA PIs with the thought that if we considered this trend as indicative, then we can expect an entire spectrum of focused to grand, parsimonious to complex, crass to elegant theories that would eventually populate the ICTD horizon, which is not necessarily a bad thing. Given the nature of theory building and theoretical discourse, all theories are inherently useful. Furthermore, the transdisciplinary nature of ICTD is leading to the convergence of disparate traditions and perspectives, resulting in synergies yet undetermined.

THE WAY FORWARD

Before the session on ICTD theory was conducted, the SIRCA researchers already had their conceptual frameworks and methodologies firmed up. Perhaps only a modicum of what I have shared could have been accommodated at that point. But if asked for a theory building agenda (which I was not); I would have given the following answer if only for consideration in succeeding proposal calls or grant rounds:

As things stand, we have merely scratched the surface of this body of knowledge that we call ICTD. We need to know more about ICTD networks and how they behave. We know that there are inherent synergies in networking but how and why do these synergies come about? We know that social networks produce social capital but we do not know when and how they do so.

As we observe the adoption of de facto software standards, the spread of mobile phones in India and China, the texting phenomenon in the Philippines, the saturation of BlackBerry units in Indonesia, and, most recently, the influence of *Facebook* on the Arab spring, we refer to the critical mass phenomenon. And yet we cannot pin down the exact point when this critical mass is achieved and the conditions that produce it.

The introduction and spread of mobile devices into the ICTD arena has become a game changer in itself. With the increasing use of mobiles in rural areas, will there still be a need for intermediaries? Telecentres? Hundred-dollar laptops? Like everything else, the strategies that we design cannot keep up with the technology that we develop.

Finally, we ask, where is all this taking us? Analysing ICTD feels like observing an unconsummated phenomenon still in the process of unfolding and about to peak. And yet we instinctively possess the conviction that there are forces at work here that we can and will, eventually and fully, understand, explain and (yes) predict.

References

Calvano, Michael. Personal Interview at the ITU Regional Office, Bangkok (2004).

Davenport, Thomas and Larry Prusak. *Working Knowledge: How Organisations Manage What They Know.* Harvard University Press, 1998.

Flor, Alexander G. *Developing Societies in the Information Age: A Critical Perspective.* Quezon City: UP Open University, 2009.

———. *ICT for Rural Livelihoods: A Scoping Study for Southeast Asia and the Pacific.* A study commissioned by IDRC-IFAD ENRAP, 2008.

———. *eDevelopment and Knowledge Management.* Los Banos: SEAMEO SEARCA, 2001.

———. *The Information Rich and the Information Poor: Two Faces of the Information Age in a Developing Country.* PhD Dissertation. Los Banos: University of the Philippines, 1986.

Freire, Paolo. *The Pedagogy of the Oppressed.* London: Penguin Books, 1970.

Habermas, Jürgen. *The Theory of Communicative Action.* Translated by Thomas McCarthy. Cambridge: Polity,1984–87.

Hayek, Friedrich. "Economics and Knowledge." *Economica.* London: Wiley, 1937.

Jussawalla, Meheroo, Donald Lamberton and Neil Karunaratne (Eds). *The Cost of Thinking.* New Jersey: Ablex Publishing, 1989.

Nonaka, Ikujiro and Hirotaka Takeuchi. *The Knowledge-Creating Company.* New York: Oxford University Press, 1995.

Porat, Marc. *The Information Economy.* Washington, DC: US Department of Commerce, 1977.

Quebral, Nora C. (1973/72). "What Do We Mean by 'Development Communication'". *International Development Review* 15 (2): 25–28.

Schiller, Herbert I. *Communication and Cultural Domination.* New York: International Arts and Sciences Press, 1976.

Schumpeter, Joseph. *Capitalism, Socialism and Democracy.* New York: Harper, 1942.

6

MESSY METHODS FOR ICT4D RESEARCH

Rahul De'

INTRODUCTION: INSIDE THE CLEAN ROOM OF RESEARCH METHODS

The semiconductor industry uses a clean room in which to manufacture silicon chips used in computers. A clean room is purged of dust particles as even a tiny mote can destroy an entire production run. A typical clean room has twelve or less particles per cubic metre, as compared to normal air that has millions of particles in the same amount of space. Those working in clean rooms have to enter through air locks that purge their clothes of particles, and also have to wear space-suit like coveralls to maintain the cleanliness.

Much of ICT4D research is criticised for not having rigorous methods — papers generalise from scanty data, case studies do not have adequate depth, statistical analysis is loose, and analytical generalisations are not convincing. The international community of ICT4D research demands rigour, in the form of tighter controls over data collection, larger and more in-depth data for analysis and the use of well-established methods, in the broad domains of qualitative or of quantitative analysis. Researchers are urged to have clear and precise plans for research, which will include setting up research questions or hypothesis based on an in-depth literature survey of the extant literature, a clearly specified plan for the kind and quantity of data that will be collected, the manner in which the data collection will proceed, and a very precise outline of the analysis method.

Call-for-proposals from granting agencies, proposal writing guidelines from research departments, and how-to books on research; all emphasise the above steps for conducting research. Everything has to be spelled out in as much detail as possible, so that reviewers, department committees, international funding agencies, and others can judge the rigour, relevance, and potential value of the proposed research. Even when the research is completed and submitted for publication in peer-reviewed journals, the reviewers want a detailed and clear exposition of the method planned and followed for the research and data collection.

The demands of rigorous research are similar to the demands of the clean room. The research has to have a dust-free plan, a mote-free manner in which it is conducted, and the researchers have to wear coveralls whilst doing their work to ensure that it remains true to the demands of the clean room. The research problem statement, data collection, analysis, model building, all have to have a clinical cleanliness about them that is up to the highest standards of the best journals in the field.

Prior Theory on Methods

The radical philosopher Feyerabend challenged the notion of scientific method in his seminal book *Against Method* (Feyerabend 1975). He argued that science does not progress by following a systematic approach of first having a theory, setting up hypotheses based on that theory, and then looking for evidence to further extend the theory or refute it. Science progresses by considering entirely new and original theories that are not tied down ideologically to older ones. Thus, scientific methods are limited when they prescribe a limited means of progressing, relying heavily on an older worldview. Progress in science is possible by an 'anything goes' method, where ideas assume dominance by some means, not necessarily because they are the best progression from an older idea.

Feyerabend's argument extends to the writing of science where the arguments, the inconsistencies, and the trace of thinking are eliminated in the final writing of the work. Theory is stripped of the history of its origin and final form.

The argument of seeking and learning in new ways is echoed by Law (2004), who says that "Parts of the world are caught in our ethnographies, our histories and our statistics. But other parts are not, or if they are then this is because they have been distorted into clarity" (Law, 2004). The world is 'distorted into clarity' when particular and rigid methods are used, and the messiness, the uncertainties, and the irregularities are removed from published works. To overcome this, new ways of seeing, seeking and understanding have to be thought of and practiced.

Problems with Clean Room Research

Much of the research in the ICT4D field cannot be conducted inside the clean room, although the attempt is made to do so. The reasons are the following:

1. Researchers use prior theory to draw up a set of research questions and hypothesis on the basis of which to conduct their research. Many researchers in ICT4D have found that this prior theory, drawn from the existing canon (the most cited papers) is usually inadequate. The extensive literature in the field of MIS, development studies, and other fields such as computer science and social studies, only touch upon the problems of ICT use and impact in society. The prior theory that is used to frame the research questions is simply not able to help fathom the depth of the situation in the field. When asking questions about design of the ICT project, its implementation, and its impact on users, the users' feelings about the technology, and so on, researchers find that the complicating contextual issues dominate their findings, and confound their research.

2. In a developing country like India, from where I'm drawing most of my examples, the issues of caste, corruption, the nature of the state, the current political economic situation, the historical processes of development, and the current economic policies, all play a role in defining the context. Researchers are able to ask focussed questions about technology use and design; however, the larger context begins to dominate the study. Researchers find that either their questions are inadequate, or that the observations they are recording, both from interviews and from field experience, hardly capture a fraction of the immense detail available to them.

3. Research insights are often obtained from observations that are made opportunistically, from sensed, rather than observed, phenomena, from reading nuances in conversations, and from the body language of respondents. For example, the power of caste hierarchies and the binds it imposes on the ordinary practices of people cannot usually be asked directly. It can be inferred from the manner in which respondents react

to questions in the presence of others, the distance that they maintain from the interviewer, and the kind of questions they avoid.

Challenges of Field Research

Some examples of the problems faced by researchers conducting ICT4D research in the field are as follows:

1. A doctoral student went to a remote district in India to study the impact of a set of telecentres that had been deployed there. He had prepared extensively to ask questions based on Sen's (2000) capabilities and freedoms model. He was able to proceed comfortably as long as his questions remained in the vicinity of information technology and its use, but as soon as he asked about issues related to the larger aspects of living in a rural, largely poor part of the state, he found he had to abandon many pre-formed lines of questioning. He had to account for the issues of caste of the respondent using the technology, the place where she lives, the time at which she walks in to access the service, the manner in which she approaches the intermediary, what aspect of her work life forces her to seek the kiosk, her social and economic position, the accent she speaks in, the manner in which she sits or stands at the kiosk, the response she receives from the intermediary, how she makes the payment for the service, whether she has to negotiate a bribe, what she has to say and hear during the interaction, and how she exits the situation (Singh 2011).

2. A research scholar (a SIRCA grantee) prepared an extensive plan for data collection regarding agricultural information in a remote rural region in India. She had prepared a questionnaire based on the existing secondary data on the kind of agriculture being practiced in that district, the demographic profile of the residents, the extent of ICT penetration, and an outline of the political and economic conditions. Upon reaching the site and after spending time there she realised that the line of questioning she had prepared was not going to help. Prior categorisation of the respondents based on secondary data and government classification were also not applicable in the region as some of these categories were non-existent or too small in number to be of significance. The real farmers there were women, and not men, as the men had migrated away to work as labour; the region was short on irrigation water; the market intermediaries, such as money lenders, had a stronger influence on the small farmers than had been stated in the literature; the respondents had very different concerns

than those of agricultural information, which they wanted to discuss with the researcher. Extension / outreach activities of the state, which the researcher was examining, were often implemented in a standalone mode with little concern or understanding of other inputs required by the farmer (Kameswari 2011).

3. Another scholar prepared extensively for data collection in two remote villages in India, one in the state of Tamil Nadu and another in Uttarakhand. The scholar had prepared structured questionnaires that had to be administered to school children. The questions dealt with the children's perception and use of computers. Apart from the issue of language (the scholar spoke a dialect of Tamil), where children repeatedly asked for clarification on certain words, the major issue turned out to be the sensitive aspect of education of girls. To quote the scholar:

> As such, questions on the need for education for girls could not be asked directly. I had to, instead, observe from interactions with teachers in the schools and with parents in the homes to have a better sense of how they viewed the girls' role and need for education, etc. I learnt that I could get a more nuanced understanding of a culturally sensitive topic through observation of these visible indications, which might not have been adequately handled if questioned directly (Ale 2011*b*).

Some Comments on Recent Publications in ICT4D

A review of three papers, recently published, and all of which were based on research funded by SIRCA, shows the dominance of clean room research. The papers report on projects in Bangladesh (Ashraf et al. 2011; Islam 2011) and India (Raman and Bawa 2011*a*) and appeared in the journal *Media Asia*. These papers were selected for this review as all three were based on field research and dealt with ICT introduction in a developing country context.

All three papers begin with a problem statement that originates in theory. The papers establish that the existing literature does not bear adequately upon the theoretical problem they are considering and hence the research is necessary and relevant. The papers then establish a theoretical frame, delving deeper into the extant theory, to identify research questions. This is followed by a detailed description of the methodology and then an analysis. As such, the papers reflect the best practices of quality academic journals in this domain.

As is also the best practice in most academic journals, the papers leave out of the discussion the mess that was part of the methodology and the subsequent analysis, and present a clean, sanitised account that leads to clear, unambiguous reasoning and durable conclusions. This is not a critique of the papers; it is why the papers are of a high enough quality to appear in a journal such as *Media Asia*.

Had the papers left in some of the challenges of data collection, such as the problems respondents had with interpreting questions, the variety of responses that were received in Bangla and Kannada and their subsequent translation into English, the nuances of body language, locale, researcher-respondent dynamics, the subtle power relations imminent in all conversations, etc., the papers would not only have been unmanageably large, but also would have not had the quality attributes respected by journals.

Finished products — such as revised journal articles — cannot really include the mess that goes into producing them. What can be hoped for is that the authors include a section on the mess that they had to clean up through the revisions and changes they had to make during the design, data collection and later analysis phases of the research.

OUTSIDE THE CLEAN ROOM

Research outside the clean room requires being opportunistic and flexible regarding the entire process of data collection and fieldwork. Well prepared plans and detailed questionnaires may have to be abandoned or changed. The manner in which interviewees are sought out, interviewed, observed and recorded as respondents has to be adjusted to the needs of the situation. In short, "messy" methods have to be adopted.

Messy Methods

So how can a researcher proceed on a messy method to conduct her research? Does it imply an abandoning of all methods, and simply doing whatever one wants? Or is there an alternative plan that could work? Doing research outside the clean room does not mean abandoning planning or preparation for doing the research. On the contrary, it requires a greater degree of preparation and sensitisation to enable the researcher to respond to the needs of the situation on the ground. In the points outlined below, we set out a possible approach to messy methods. These may not include all possible scenarios that may arise in the field, as they possibly cannot, but they do point to a number of ways in which the researcher can prepare.

1. Prior theory can help in identifying unanswered questions and also the ways in which research has been conducted in the past. Researchers have to uncover the theoretical, but more importantly, the methodological gaps

in the literature. It is important to read published research to examine
the ways in which the study was conducted, and identify what could be
the issues that remained uncovered, owing to the method used.

2. The research questions identified by the researcher, to guide the study, must
   remain flexible. They should be revised based on the initial observations
   and responses obtained from the field. The changes thus introduced must
   be recorded, along with a set of reasons for why this was done.

For example, one researcher prepared a structured questionnaire based on a
well-established scale, called the Smiley scale, for obtaining responses from
children. In the field, the researcher found that the children simply did not
understand the scale:

> Children did not understand what the Smiley scale meant or how to
> make sense of them. Although the corresponding meanings of the smiley
> were reflected in the scale, children got distracted with the faces and text.
> Moreover, despite repeated explanations, children did not know what
> it meant by the degrees in which to agree or disagree. ... As a result of
> these drawbacks, changes were made to the scale and items tested. The
> components of the five-point scale Smiley-face scale were removed ... [the
> revised scale] had a question on one side (e.g. I can use a laptop) and the
> same question in reverse order (e.g. I cannot use a laptop) on the other
> side of the scale. This made it easier for children... (Ale 2011*a*).

Other researchers (SIRCA grantees) too had to revise their questionnaires:

> We had prepared a topic guide which contained broad themes around
> which we asked questions to different respondents. We revised this topic
> guide about two times after the initial preparation. The revisions were
> on the basis of the findings and insights we got from the field and the
> emergent themes that we needed to ask questions around to nuance some
> of our contentions and hunches (Raman and Bawa 2011b).

The examples show that researchers do respond to the needs of the ground and
modify their methods. It would be highly instructive if all ICT4D research
reflected these adjustments and recorded the rationale for them.

3. The task of identifying potential respondents, their profile and their
   geographic location is a challenge. This has to be a rough cut outline,
   the exact set of respondents can be selected from the location chosen,
   however it is usually difficult to be very precise, at the outset, of the
   profile needed. It is very important to read about, as much as possible,

the sociological aspects of the respondents and be sensitive to their social position and status, their economic status, their caste affiliation, their religious affiliation and practices, their community and network, their work practices, and so on. These sensitivities will guide the researcher as to the nature of the respondents and the nature of responses they are receiving.

4. The researcher has to also closely examine his or her own subject position with regard to the respondents. Are they approaching the respondents from a visible and easily marked position of power? Is it easy to see that they belong to a dominant caste (as is usually the case in India) with regard to the respondents? If the researchers are using interpreters, are they able to gauge the confidence or distance the respondents have to the researcher? The researchers must record their interactions in detail, revealing their own subject positions as carefully as possible.

5. The issue of ethics arises in the context of ICT4D research (Traxler, this volume). When researchers approach members of poor and marginal communities, the implied reason is usually not evident to the respondents. For example, some respondents, when approached for a survey for a project in India, assumed the interviewer was a government functionary. They responded eagerly, and answered questions in detail, with the hope that they would find some redress to the woes they were narrating. The researcher was in a moral bind to assure them that he was not a government employee and had no powers to take any action.

   The challenging question that arises here is whether the researcher should have withdrawn knowing fully-well that the respondents had the wrong impression, or whether he should have continued with the deception to obtain the data. Though there are no easy answers to this question, the researcher should at least record such conditions and situations in which they arise.

6. The hidden and subtle aspects of life in rural India, particularly for marginal populations, have to do with caste power and practices, the power and violence of state functionaries, the problems of corruption and abuse of power, the role of shifting political alignments, the extent of the state's visibility, and so on. These aspects can be asked about directly to respondents, but in many cases they have to be asked in a subtle manner and possibly not asked at all. For example, one researcher found that it was possible for a dominant caste farm owner to give a missed-call on a mobile phone to a non-dominant caste landless employee, but the employee could not do the same, i.e. give a return missed-call. The researcher sensed that this was an issue of caste power and hierarchy but could not

probe further as the employee declined to respond. It was not possible to raise this issue with the farmer either, as it may have had negative consequences for the employee. The researcher's response was to note this aspect of technology use and ask others to verify it, using questions that could explain the underlying reasons for the phenomena.

7. Messy methods must account for and include ways of sensing and recording data that are unusual, opportunistic and transitory. A researcher has to be conscious that not all the data will fit the answers to the questions being asked, but may point to issues that were not within the scope of the research. The researcher has to respond to such data and explain why they can or cannot be included within the analysis.

CONCLUSION

It is imperative that ICT4D research that examines issues in complex settings, such as those in rural India, explore in detail and expose the challenges that the data collection method posed, the limitations of the prior theory used, and the adaptations that the research had to resort to. This requires training in and sensitivity to the nuances of data collection, and the ability to recognise that which the researcher is not able to observe and record readily.

ICT4D research will do well to step outside the clean room and acknowledge the world with all its mess and complexity. Journal reviewers and editors, conference editors, funding programme managers and academics and practitioners have to acknowledge and encourage messy methods. New theory building will require a new epistemology.

Acknowledging messy methods of research implies acknowledging that practice is also messy. Designing, constructing and deploying information and communication systems in rural and remote communities that will affect the lives of marginal populations requires dealing with unplanned events, unexpected situations, contingencies and situations and locales far beyond that envisaged. ICT4D research does acknowledge these challenges, but more as an afterthought rather than an integral part of the report. Adopting messy methods will possibly also induce practitioners to acknowledge that practice in ICT4D too is done outside the clean room.

References

Ale, Komathi. "Cheap and good? An ICT in education intervention to evaluate the impact of low-cost computers on the self-efficacy and literacy of primary school children in rural India." Master's thesis, Nanyang Technological University, Singapore, 2011a.

————. Personal communication. (2011*b*).

Ashraf, Md Mahfuz, Helena Grunfeld, Roger Harris, Md. Nabid Alam, Sanjida Ferdousi, and Bushra Tahseen Malik. "An explorative study of ICT for developmental impact in rural areas of Bangladesh." *Media Asia* 38 (2011), 22–31.

Feyerabend, Paul. *Against Method*. London: Verso, 1975.

Islam, M. Sirajul. "Evaluation of an m-service for farmers in a developing region: A case study from rural Bangladesh." *Media Asia* 38 (2011), 41–51.

Kameswari, Vyakaranam. Personal communication. (2011).

Law, John. *After Method: Mess in Social Science Research*. United States: Taylor and Francis, 2004.

Raman, Bhuvaneswari and Zainab Bawa. "Interacting with the state via information and communication technologies: The case of Nemmadi Kendras in Karnataka." *Media Asia* 38 (2011*a*), 52–62.

————. Personal communication. (2011*b*).

Sen, A. *Development as Freedom*. New Delhi: Oxford University Press, 2000.

Singh, Jang Bahadur. Personal communication. (2011).

7

ETHICS AND ICTD RESEARCH

John Traxler

INTRODUCTION

The ethics of ICTD research are important and challenging for both emergent and established researchers. The reasons for the importance of research ethics are simple and obvious but still worth enumerating. Basically, research that does not conform to explicit ethical guidelines may be:

1. seen as improper or immoral
   a. attracting opprobrium and condemnation
2. breaking laws or regulations
   a. attracting litigation and prosecution
3. unacceptable to the research community
   a. affecting funding, publication and promotion

This chapter reviews the nature of the challenges facing ICTD researchers seeking to work ethically, firstly from the conventional institutional perspective and then from community perspectives including those of online communities.

ETHICS: THE BACKGROUND

Before we look at ethics in the context of research and of ICTD, we need to review some basic and accepted ideas. According to Farrow (2011, p. 102) ethics is, "developing a systematic understanding of why particular behaviours are (or should be) considered right or wrong." This sounds simple, clear and obvious. However, "Ethical concepts are slippery and complicated, and for many it's natural to lapse into either a kind of lazy ethical relativism ('follow your own path') or to conform to 'the rules' or mores as we find them" (2011, p. 103). "Furthermore, the diversity of mobile devices and their contexts of application can make it very difficult to anticipate and make a judgement about ethical issues that might arise. This means that the advice given in general ethical guidance is often vague and non-transferable (2011, p. 104). This is not just a function of the rapidity of technical change — Farrow is specifically talking about mobile technologies but his remarks are true of all ICTs — but also of the rapidity of appropriation by cultures and sub-cultures.

Farrow provides a basic and accessible classification of different ethical (or rather 'meta-ethical', the study of the meaning and use of moral language) perspectives, namely the deontological, the consequentialist and the virtue-based, as follows,

> "Deontologists are interested in the precise nature of our moral obligations, where conflicts between them may exist, and where there might be exceptions." ... "A deontological perspective may be more useful when considering the kinds of responsibilities and duties that are relevant to particular m-learning scenarios." (p. 104)

> "...any consequentialist meta-ethics assess the 'rightness' or 'wrongness' of actions specifically in terms of their consequences: consequentialists believe that talking about the rightness or wrongness of an action is equivalent to talking about the desirability of the (likely) outcomes" (p. 104).

Farrow feels that, "most institutional policies are normally drawn up with broadly utilitarian principles or methods of justification in mind" (p. 104). Utilitarianism is the most common form of consequentialism and offers a way to evaluate decisions and outcomes in terms of their aggregate effect or 'utility'. It could be understood as a kind of cost/benefit analysis of the different potential outcomes of an action and might also relate to the position that ends justify means.

> "A virtue ethics approach focuses upon the desirability of traits, skills and characteristics of agents: virtue ethicists believe that ethics is about cultivating the qualities and habits that contribute to a good or 'flourishing' life." (p. 105)

To summarise, "The deontologist believes that ethics is a matter of duty and responsibility. The consequentialist believes that ethics is a matter of promoting good outcomes. The virtue ethicist believes that ethics is about cultivating good habits in order to flourish." (Farrow 2011, p. 106)

This is a helpful and apparently universally relevant classification but one that is clearly, albeit implicitly, based in a European, specifically a rational modernist European world view, in the people, history and culture of Western Europe. It may have subsequently fragmented within academic philosophical circles and been problematised by various authors who are characterised as *postmodern* — at the same time and for the same reasons as the underlying epistemologies (and research methods) of Western science have also been questioned — but it is still the prevalent ideology in those countries and institutions around the world that have been influenced by European ideas. Beyond this mainstream and arguably global view, it is worth briefly mentioning and explaining the postmodern perspective. This is, in fact, quite difficult because there is no unified postmodern perspective as such, only various individuals taking positions that follow from the modern, from modernity.

We should recognise that, "Research located in the contested intellectual space that is 'development' needs to be able to answer the fundamental question of what is understood as development" (Kleine 2009). This is not only a significant and fundamental question for the practice of *development* in general but also to its ethical dimension in particular. The different ethical positions clearly colour why people, organisations and communities support development but also colour their objectives and goals; only the consequentialist, or rather the utilitarian position seems to actually identify and measure these, as the utilitarian talks about *the greatest good for the greatest number*. Even this becomes problematic once we ask about timescales or move away from purely economic metrics such as GDP per capita.

Some Examples

Perhaps in order to get some sense of the issues, we should look briefly at some potentially problematic scenarios, drawn from working with mobiles to support learning in developing regions of Africa. Consider the following hypothetic examples written in the first person:

1. We plan to conduct a needs analysis prior to implementing mobile learning for rural farmers by going into communities and using informal focus groups to ask community members, amongst other topics, whether they own and use, or have access to their own mobile phones, and if not, why.

2. We need to run a pilot study of the impact of mobiles on learning in a typical rural primary school. We will issue all the children with basic mobile phones for a period of six months with inclusive airtime and then retrieve the phones on the final evaluation visit.

3. Our pilot project to deploy mobile phones with maths apps in four rural primary schools now has the backing and support of the ministry of education and the provincial education department. The project will be started and the phones will be issued to children as soon as possible.

4. The class teachers for the project have been nominated by their head teachers and will have been trained at a national residential event. Travel and accommodation will have been met by the funders and the teachers will receive per diems.

5. As part of this project, we will issue the class teachers with a loaned laptop and drawn a legal indemnity agreement assigning responsibility to them for them to sign on the day we visit their respective schools.

6. We can use each of these visits to capture some background video of the teachers, and their classes using the mobiles for subsequent presentations and reports.

7. Gathering data from teachers supporting the rural primary school project, and also head teachers and administrators, will require several days of their time and it is proposed that we recompense them with a certificate and academic credit.

8. As part of the assessment for their ongoing professional development, practising mid-wives will be required to use camera-phones to capture short video clips of their practice. These will be anonymised and only viewed by their tutors.

9. An international funder is just about to issue a call to institutions in the 'developing' world asking for research proposals in mobile learning. The call makes no reference to ethics procedures.

10. A high profile pilot has established that high specification feature phones greatly enhance the self-efficacy and self-esteem of teachers in the field. There is predictably no funding to scale or sustain the project.

11. A negative evaluation of a national pilot project looks likely to jeopardise the idea of using mobiles to support in-service teacher training. The evaluators are proposing to restrict access to their report.

12. In order to give a small and remote community greater access to national job and education opportunities, we propose to give them airtime and access to mobile-based English language courses. Their language has less than 600 speakers and is listed by UNESCO as 'endangered'.

ETHICS WITHIN THE INSTITUTION

Professional academic researchers are expected to be familiar with their institution's ethical procedures, though these may vary dramatically from institution to institution, especially when we compare such procedures across different countries and cultures. These procedures draw on a vague and tacit acceptance and blurring of the perspectives outlined above.

Their explicit and over-arching concern in theory is to ensure that the proposed research *does no harm*. In practice this will embrace a fairly stable and conventional list of issues including:

1. Informed consent: the principle that human participants in research have understood and consented to taking part in the research activity. It is the key ethical issue from which in effect all the others flow. Whilst simple in principle, working with novel, abstract and complex technologies and working with different/distant age groups, communities and cultures pose enormous and subtle challenges (Traxler 2008; Lally 2012). The subsequent principles, and others, could be argued to be subservient to informed consent in the sense that they are potentially negotiable within the consent, as long as the consent also stays within the law. The law itself however may not be simple or clear in areas of innovative virtual technologies.

2. Confidentiality and anonymity: the principles established and agreed about how research data is shared, disseminated, discussed and published. Describing and understanding the relationships, visibility and permanence of online and published data can be complex and unpredictable, as can the concepts of identity and privacy.

3. Data protection: the principles, often enshrined in statute, that cover the keeping, sharing, disclosing and finally destroying of research data.

4. Liability and responsibility: the principles that describe what harm and risks might be associated with the research and who or what is responsible or liable, and the nature and extent of that liability.

5. Compensation and reward: the principles that articulate how and how much participants in research are to be rewarded or recompensed — two rather different concepts — for their participation. Aside from the issue of skewing the sample or the responses, this principle can be difficult to operationalise across culture divisions where different value systems or simply different economic standards obtain.

6. Briefing and debriefing: the principle of ensuring that participants fully understand the nature of their participation beforehand and of ensuring

that they are fully counselled or comforted afterwards. The former is problematic in potentially prompting or biasing responses.

The procedures will often require some form of institutional clearance before research can commence, and this will come from an institutional committee or board.

Institutional review boards (IRBs), by whatever name, are usually composed of senior or middle academics representing various disciplines or divisions from across their constituency. They may have junior or preparatory sub-boards at a departmental or discipline level; they will report to an institutional body, perhaps the institutional governing body or research committee. They usually operate by requiring intending researchers to obtain clearance, based on filling in standard forms, before any research can commence. The forms may be more or less detailed, asking about research instruments and procedures, and then using these to classify the proposed research according to the perceived likelihood and nature of risk and harm. According to these criteria, the proposed research will be categorised and escalated to a level that feels competent to reach a judgment. Researchers may receive a permission to proceed; a request for further details or a request for modifications. A common problem is that of forcing researchers prematurely into details and specifics before perhaps a literature review or an initial empirical investigation has taken place or before project staff have been hired or selected. Furthermore, funding for the proposed research may have assumed particular activities in the expectation of ethics clearance which in the event may not be immediately forthcoming.

Furthermore, many professional researchers feel themselves subject to the ethical codes or guidelines of a professional body such as the Association of Internet Researchers (AoIR). These codes, guidelines or frameworks may be quite prescriptive and directive, and specific to the limited context of the research profession or they may be formulated in terms of underlying moral aspirations and of the roles and responsibilities of the research profession in a wider social context. Amongst the professional bodies there has in the last couple of decades been a move away from prescriptive codes to frameworks that ask professionals to acknowledge and perhaps reconcile underlying principles such as beneficence (doing good), non-maleficence (avoiding harm); autonomy (respecting choice) and justice (equality of access to resource) (Wishart 2009).

These various arrangements may all however be poorly suited to research such as ICTD research, working at the interface of technology and society, especially working with rapidly evolving technologies being rapidly appropriated and working with diverse societies (and cultures, sub-cultures and counter-cultures) distant and distant from the institution.

Institutional ethics procedures and specifically IRBs and their various instantiations nationally are likely to be operationally inadequate for our research work in ICTD because:

1. They sometimes focus on issues outside a purely ethical remit, for example worrying about the public relations impact of a piece of research or concerns about a proposed methodology, sample size and research instruments, for example;
2. They may have only a limited and untrained knowledge of the law but nevertheless may be preoccupied by potential legal issues;
3. They may be composed of researchers from inappropriate disciplines, philosophies or backgrounds, for example from animal biology or human psychology, from a purely quantitative practice or from empiricist or positivist philosophies; even within the ICTD research community, the prevalent upbeat positivist techno-deterministic stance adopted by many researchers may be at odds with more innovative approaches;
4. They may not be technologically up-to-date or critically aware. We should note in passing how the complexity of technology can lead to unfortunate and unexpected consequences (Wishart 2009);
5. They may have limited knowledge of other cultures or countries; they may be composed largely of middle-ranking career academics, not necessarily research-active;
6. They may operate prescriptive and inflexible codes or formal administrative procedures; and
7. They may be perceived as paternalistic or as a way of individual researchers to abrogating personal responsibility; they may be seen as a formality involving merely compliance.

These observations are obviously often valid for other disciplines too.

Research projects and research staff specifically in ICTD can sometimes have other ethical challenges:

1. Their institution may actually have no ethical procedures, or only ones intended to address minimal legal compliance.
2. They may be part of international consortia, with unclear or overlapping national jurisdictions or practices.
3. They may be subject to competing or overlapping ethical codes or procedures, from for example professional bodies and employers.

4. They may be working in consortia, again subject to competing codes or to none.
5. They may be funded by donors, corporations or governments for whom ethical practice is not a clearly articulated priority.
6. Sometimes a project is represented, not as a research project and subject to research ethics procedures, but as an intervention or a deployment or a mere technical change, and not subject to research ethics procedures.

In terms of some of ICTD's constituent disciplines, in the social sciences, any covert methods are obviously problematic; in internet research, anonymity and confidentiality are often problematic since sources, attributions and identities are relatively easy to uncover and connections easy to make and in research in distant and different communities, researchers and their participants may have very disparate ideas about what constitutes fair recompense for participation.

Some ICTD interventions, either research or deployment, outside the university sector, undertaken by governments, agencies, foundations and universities, have no apparent mechanism, however simple, for periodically asking the question, "Might our actions or words cause harm?" and then the more subtle question, "What in the eyes of the people we are working with would constitute harm?" One failing of many institutional procedures is that they do, in fact, ask the first question but only once, before any interactions in the field. Also researchers might not be familiar with the idea of risk analysis and the procedures to quantify and respond to risk in their work.

This is however what one might call a technical or tactical analysis of research ethics in ICTD, an analysis suggesting that the researcher's institution has the sole right to decide what constitutes harm in the communities where ICTD researchers do their work and that modifications or improvements in procedures can still make the institutional arrangements appropriate, adequate and acceptable. A different perspective, the perspective of the communities *with* whom ICTD researchers work, perhaps that of the communities *for* whom ICTD researchers work, challenges this analysis.

ETHICS IN THE DEVELOPING COMMUNITY

Within any society, community or culture, there are likely to be norms, ones that uniquely characterise that society, community or culture. These will place that society, community or culture on a range of axes defined by, for example:

1. Risk-taking *vs.* risk-avoidance
2. Individualism *vs.* collectivism
3. Hierarchy *vs.* equality
4. The extent of gender inequality
5. Control *vs.* consensus, following Hofstede (2001) and also,
6. Innovativeness *vs.* conservatism, according to Rogers (2003), though probably not independent of the risk axis

Furthermore, any society, community or culture is also going to be uniquely characterised by a specific balance between the formal, the established and the institutional on the one hand and the informal, the indigenous, the local and the vernacular on the other, and perhaps the counter-cultural, the subversive and the disruptive too, amongst peer groups, communities, families, kinship groups and elders. So from the perspective of outsiders, of researchers coming from outside, there is much potentially that might be *not worth mentioning* or *taken for granted* (Rugg and McGeorge 1999), so many areas where harm might be inadvertently perpetrated. The earlier account of meta-ethics should perhaps be seen in this wider cultural context.

These remarks apply with varying granularity and relevance to communities from nations down to villages and they correlate or explain differences in ethics across different communities, ethics in the shared sense of what is approved, acceptable, appropriate, allowable or permissible in terms of interaction, relationship, manners, exchange, humour, posture, language, discourse, fashion and behaviour in any community. Of course, individuals often belong to more than one community and probably aspire to be accepted by several others. Looking at the basis of the ethics for any community exposes the complexity confronting institutional procedures that attempt to regulate research interventions distant from the institutional culture and community.

In South Asia, as in some other parts of the world, countries and regions are characterised by a multiplicity of religions existing side-by-side, further contributing to the ethical complexity and granularity. These religions form the foundation of their communities' ideas about ethics and morality but differ each from the other. Furthermore in South Asia the rapidity of technological adoption, change and appropriation leads to generational differences, not always based on chronological age.

In general, in ICTD research, there is a significant distance or difference between the researchers and what we might call, the *researched* (there is no neutral terminology: *subjects, respondents, co-researchers, stakeholders* are all terms that betray particular ideological perspectives). It may be a difference in geographical or cultural origin; it will probably be a difference in socioeconomic class, perhaps in ethnicity, religious affiliation and probably in education.

To take this further, Richard Heeks (2008) implicitly separates ICTD research from mainstream empirical research by identifying an agenda in terms of the poor. He theorises, in discussing his idea that ICTD moves in generations, from ICT4D 0.0 to ICT4D 1.0 and now to ICT4D 2.0, that "ICT4D 2.0 focuses on reframing the poor. Where ICT4D 1.0 marginalised them, allowing a supply-driven focus, ICT4D 2.0 centralises them, creating a demand-driven focus. Where ICT4D 1.0 — fortified by *the bottom of the pyramid* concept — characterised the poor largely as passive consumers or recipients, ICT4D 2.0 sees them as active producers and innovators." Within ICT4D 2.0 he sees:

1. *Pro-poor* innovation occurs outside poor communities, but on their behalf (p. 29).
2. *Para-poor* innovation is done working alongside poor communities (p. 30).
3. *Per-poor* innovation occurs within and by poor communities (p. 30).

This classification must logically entail different ICTD research ethics, ranging from *pro-poor ethics*, that is, on their behalf, from institutions and IRBs downwards in the way that we have described, to *per-poor ethics*, the ethics from within the communities. In making this inference, we move towards the methodological positions of many progressive ICTD researchers. They espouse participative methods or user-centred research methods and the obvious corollary must be participative and user-centred research ethics, or perhaps *user-generated ethics*, entirely different from research ethics originating within distant institutions. The challenge is now to turn this principle into practicality.

ETHICS IN ONLINE COMMUNITIES

We move now from research ethics based within real communities to research ethics within online communities. Increasingly in ICTD, the focus for pragmatism and for innovation is shifting away from static, centralised, institutional ICT technologies and shifting towards popular, personal ICTs. These include:

1. Web 2.0 technologies such as blogs, wikis, mash-ups, podcasts, RSS feeds and user-generated content;

2. Mobile connected personal devices, epitomised by the *Smartphone*;
3. Social networks such as *Facebook* and *LinkedIn*;
4. Micro-blogging sites such as *Twitter* and *Mxit*;
5. Cloud-based data and communications services such *as Gmail, Drop-box, Flickr, YouTube, GoogleGroups, Wikipedia, Evernote*;
6. Voice-over-IP, epitomised by *Skype*;
7. Retailing services such as *iTunes, Amazon, Android Market,* and perhaps also in the most metropolitan areas;
8. Immersive virtual environments like *Second Life* or *Habbo Hotel*; and
9. Online multi-user gaming

These are broad and not mutually exclusive categories, and the emphasis in ICTD is usually on only a limited subset. What is common to all of them is however the increased level of *ownership*, personalisation, agency, control, familiarity and confidence that they give to individual users and informal groups, unmediated by organisations and institutions (albeit still controlled by large multinational corporations in quite a few cases and funded by ever-more sophisticated and targeted advertising). The balance of take up and engagement amongst these various categories will depend on local conditions such as connectivity, buildings and mains services, tariff regimes and network coverage, hardware costs, availability and accessibility, as well as cultural and political factors.

These technologies are creating more and more places and modes that people can inhabit, where communities can form, where opinions, ideas, images and information can be produced, stored, shared, evaluated, transmitted, consumed and discussed. Each community, no matter how informal or ephemeral, will have expectations about language, manners, humour, posture, taste, fashion, etiquette, gesture and behaviour, about what is admirable, acceptable, appropriate and allowable. Essentially each community will have its own ethics and standards. Directly or indirectly, these ethics and standards that each community defines for itself define also what constitute harm, oppression, embarrassment, hurt, oppression and shame for that community.

As an aside, it might be relevant to note that attempts have been made to link mobiles, mobility, post-modernity and *development* (Traxler 2008). This linkage, however weak, does have implications for ethics. We have already made the points that *development* research and mobiles in societies have ethical aspects. Both can be linked to philosophical positions that move beyond the established traditions of empiricism and logical positivism. These positions are complex and confused but one of them, for example, characterises our societies as moving into *liquid modernity* (Bauman 2000) — apocryphally paraphrased as *permanent beta*. Modernism might be crudely characterised as the mind-set and acceptance that history and humanity are going somewhere, probably somewhere good; that language and other symbols can describe reality (and that reality as a objective, shared, consistent and unambiguous source of all our experiences actually exists); that both cause and effect and also right and wrong are simple and stable; that reason, science and education are benign agents of change and improvement. These are a handful of modernism's foundational *grand narratives* (Lyotard 1979). Other derivative ones might include Darwinian evolution, Marxist accounts of history, Freudian psychoanalysis, the scientific method as a mechanism for establishing truth and the ideal of the nation-state. Development is probably one of modernism's lesser narratives but nevertheless one that is widely held and one that justifies European interventions in other cultures from the 17th century onwards. Postmodernism as we said earlier can only rigorously be defined as anything and everything that comes after modernity. Mobility, specifically the mobility and connectedness afforded by mobile phones, can be argued to change or challenge so many aspects of different cultures, particularly the solidity of our knowledge, identities, cultures and institutions, as to take beyond the certainties of modernism.

We have talked about the ethics of ICTD in terms of the ethics of ICT, saying that now increasingly ICT supports meaningful communities in various online spaces, and in terms of the ethics of *development*, where we engage with *real* communities but ones very different or distant from those of academic researchers. Curiously when commentators tell us that *Facebook* or *Twitter* is the *nth* biggest country in the world, they are inadvertently making the point that links these various conceptions of country or community. Yes, *Facebook* and *Twitter* are different countries, and "they do things differently there" (Hartley 2002). Similarly, when commentators talk about the *cyberspace*, *blogosphere*, the *twitterati* and *phonespace* full of *digital natives*, they identify (however ironically or tentatively) broad and relatively inclusive online *countries* within which many smaller communities exist populated by individuals with various affiliations and ever-more complex *digital identities*.

DISCUSSION AND RECOMMENDATIONS

So we come back to the question of how researchers, universities and agencies, amongst others, can act in ways that are ethically acceptable to communities outside their own, and furthermore, we come to the point that researchers cannot hope to operate sustainably and credibly unless they do act in ways which are aligned and acceptable to these communities. By the nature of the question, there can be no universal prescription in ICTD research and interventions. Perhaps part of the answer is to formally involve community members in the ethics process; in an informal discussion, part of background research in order to support English language teaching in South African primary schools, I had to explain and apologise for the possibility of *spam* in our proposed messaging system. Surprisingly, the teachers were keen for as much *spam* as possible. Results like this might reinforce the proposal to involve participants at an early stage. Community understanding of sophisticated and abstract technology systems will necessarily grow and evolve as projects progress so there is also an argument for a continued community involvement in ethics procedures.[1]

Research ethics is probably an area where globally different research communities are all less than completely competent and adequate, and mostly not up-to-date either and so research ethics is an area with potential for considerable peer-group, international and inter-disciplinary capacity building. It would be a mistake to assume that early career researchers in ICTD *just* need training. The research communities express themselves through their conferences and publications with individual editors and reviewers, each acting as gatekeepers according to their own preferences, priorities and capabilities. This can easily mean that a conference or a publication, for example in this case *Media Asia*, has no consistent policy or threshold in terms of research ethics.

In projects based in institutions with an under-developed research culture or an under-developed research management, there is a risk that there will be no ethics oversight; in projects funded by one organisation and executed in another, there is a risk that each will assume the other is exercising ethics oversight; in projects with commercial or corporate partners, there may be little sympathy for the overhead of ethical oversight. These are cogent reasons why organisations at the top of the funding ecology ought to take it upon themselves to tell their projects, partners and clients how ethics oversight will work. If they do not, there is an established risk that no one else will. This could in practice be quite simple, perhaps insisting that a robust and appropriate procedure become a condition of funding the programmes and initiatives, and that this procedure is refined and operationalised as the funding works its way further down the research ecology to constituent projects and individual researchers.

Note

1. For a discussion of other aspects of consent as a process rather than an event, see for example, "Informed Consent and the Research Process: An ESRC Research Methods Programme project" online at www.sociology.soton.ac.uk/Proj/Informed_Consent/debates.rtf.

References

Banerjee, Indrajit and Stephen Logan (Eds.). *Asian Communication Handbook 2008.* Singapore: Asian Media Information and Communication Centre, Friedrich Ebert Stiftung and the Nanyang Technological University, 2008.

Bauman, Zygmunt. *Liquid Modernity,* Cambridge: Polity, 2000.

Claude Shannon (1948). "A Mathematical Theory of Communication". *Bell System Technical Journal* 27 (July and October): pp. 379, 423, 623–56.

Farrow, Robert. "Mobile Learning: A Meta-Ethical Taxonomy." *Proceedings of Mobile Learning IADIS,* 10-13 March 2011, 102-110. Avila, Spain: IADIS.

Hartley, Leslie Poles. *The Go-Between.* New York: NYRB Classics, 2002.

Heeks, Richard. "ICT4D 2.0: The Next Phase of Applying ICT for International Development." *Computer* 41 (2008): 26-33.

Hofstede, Geert. *Culture's consequences: comparing values, behaviours, institutions and organisations across nations.* Thousand Oaks, Calif.: Sage Publications, 2001.

Kleine, Dorothea. "ICT4What? — Using the Choice Framework to Operationalize the Capability Approach to Development." *Proceedings of ICTD2009,* 17–19 April 2009, 108–17. Doha, Qatar: Carnegie Mellon.

Lally, V. "Researching the Ethical Dimensions of Mobile, Ubiquitous, and Immersive Technology Enhanced Learning (MUITEL): A Thematic Review and Dialogue," *Interactive Learning Environments* 19, no. 6 (June 2012).

Lyotard, Jean Francois. *The Postmodern Condition: A Report on Knowledge,* 1979. <http://www.idehist.uu.se/distans/ilmh/pm/lyotard-introd.htm>

Rogers, Everett. *Diffusion of Innovations.* New York: Free Press, 2003.

Rugg, Gordon and Peter McGeorge. *Questioning Methodology.* Northampton: University College Northampton Faculty of Management and Business, 1999.

Traxler, John. "Mobility, Modernity, Development." *Proceedings of 1ˢᵗ International Conference on M4D Mobile Communication for Development,* 11–12 December 2008, edited by J. S. Pettersson (2009). Karlstadt, Sweden: Karlstadt University Studies.

Wishart, Jocelyn. "Ethical considerations in implementing mobile learning in the workplace." In *Mobile Learning: Transforming the Delivery of Education and* Training, edited by Mohamed Ally. Athabasca: University of Athabasca Press, 2009.

8

ICTD CURRICULUM DEVELOPMENT AND PROFESSIONAL TRAINING
Mainstreaming SIRCA Research Models

Alexander Flor and Roger Harris

RATIONALE

Strengthening Asian ICTD research capacities goes beyond research and mentoring processes. In Asia as in the rest of the world, large and small scale ICTD undertakings at the regional, national and community levels are being planned, implemented and evaluated by international development assistance institutions, bilateral aid agencies, national and local governments, as well as the private sector. These programmes and projects have built-in research components that require adequately trained ICTD managers and staff.

The SIRCA programme can directly contribute to this need in two ways. Firstly, its research results should add to the content of ICTD instruction. Secondly, its research models should be incorporated into education and training. When linked to curricular content and professional training, the impact of SIRCA may assume a multiplier effect perhaps equal to if not more than the publication of its research findings because of its influence on development policy and practice.

OVERVIEW OF ICTD EDUCATION IN ASIA

The impetus for this chapter came from an observation of the lack of adequately trained ICTD manpower in Asia because of the dearth of quality formal and non-formal training programmes in this area. There may be a glut of information technology (IT) programmes, a respectable number of communication science programmes and quite a few development studies programmes in Asian educational and training institutions. However, programmes that combine IT, communication science and development studies that adequately capture the synergies inherent in their convergence are few and far between.

Information Technology Programmes. Asia's prominence in IT services outsourcing has spawned a wide array of programmes offered by both public and private institutions to supply industry manpower requirements. These programmes include baccalaureate and graduate offerings on information science, information technology, information systems, computer science, computer engineering and software engineering. They cover the entire spectrum of hardware, software, networking and systems management concerns of IT.

Communication Programmes. Similarly, several Asian academic institutions have strong programmes on the communication arts and sciences thanks to the region's robust media, entertainment and advertising environment. A long list of communication schools from South to Southeast Asia offer baccalaureate, masters and doctorate programmes on communication science and communication arts (Asia Communication Handbook 2008).

Development Studies. Additionally, networks of South and Southeast Asian state colleges and universities offer academic programmes on rural sociology, community development and development studies. Generally, South Asian curricula are patterned after the development studies and economics programmes of UK (Sussex and Manchester). These include the University of Delhi, the University of Colombo, the University of Punjab and the University of Dhaka. Southeast Asian curricula are modelled after the rural sociology, community development and agricultural economics offerings of American land grant universities. Examples of the latter are: the University of the Philippines; Kasetsart University in Bangkok; Universitas Pertanian Bogor; Gadjamada University in Yogjakarta; and Universiti Putra Malaysia.

ICTD Courses. There may be a few universities that offer individual courses titled information and communication technology for development. The Wee Kim Wee School of Communication and Information (WKWSCI), of the Nanyang Technological University (NTU) in Singapore offers a graduate course on ICTD whilst UPOU, the open campus of the University of the Philippines system, offers two undergraduate courses titled Information and Communication for Development (ICT4D) and Knowledge Management for Development (KM4D). The University of the Philippines Los Baños offers graduate and undergraduate course on telecommunications for development. Nevertheless, a full-fledged ICTD programme at the graduate or undergraduate levels that integrates information technology, communication science and development studies is yet to be established by any one of these universities.

DYNAMICS WITHIN ACADEMIA

Traditionally, information technology, communication science, and development studies belong to different faculties within any given institution of higher learning. For so long, information technology has been regarded as the domain of engineers, applied physicists and mathematicians. Communication science, on the other hand, has been perceived as falling within the purview of the social sciences and humanities. Such is the case even if educationists agree that the study of technology should not be divorced from the study of people. In fact, information technology and communication science share similar if not identical roots. Specifically, the 1948 Claude Shannon and Warren Weaver paper, *The Mathematical Theory of Communication*, presented a model of communication that later served as the basis for the now standard source-message-channel-receiver-effect (SMCRE) model. Every communication professor, researcher or student is conversant with the elements (and relationships amongst the elements) of the Shannon and Weaver Model. However, few are aware that in the same paper, the authors seminally introduced the concept of the *bit*, the basic unit of information, which was to become one of the most basic concepts of information technology. Since then, the information and communication sciences have developed separately, the former assuming a mathematic-logical paradigm and the latter adopting a socio-psychological approach.

Considering the dynamics within the academe, it is only expected that a two-way or three-way convergence rarely takes place. Like their counterparts in the West, Asian universities develop their teaching curricula from specialised academic departments that would endorse these to faculties or colleges for approval by a university council. In the process, 'turfing' becomes a real concern. Curricular offerings tend to be structured along established disciplinal lines instead of multi-disciplinary, inter-disciplinary or, more appropriately in the case of ICTD, transdisciplinary areas.

CURRICULAR MODELS

A global search for a model curriculum on ICTD would point towards the Master of Science programme on ICT for Development offered by the School of Environment and Development of the University of Manchester with Richard Heeks as Programme Director. The programme is a combination of four compulsory core courses on ICT and development management, four electives on related topics, credited overseas fieldwork and agency research visits, and a 12,000 to 15,000-word dissertation.

In Asia, there may be a couple of related curricular models that can serve as exemplars for an ICTD programme: the development communication curriculum of the University of the Philippines Los Baños; and the multimedia studies curriculum of the UP Open University. One deals with the convergence of communication science and development studies. The other combines information technology and communication science.

Development Communication Model. The development communication curriculum is perhaps the closest model upon which an envisaged ICTD programme may be patterned after. It combines communication science with development studies as well as technical courses associated with specific development sectors (i.e., agriculture, forestry, environment, human ecology, etc.). This curricular model was developed in Los Baños in the early 70s and has since been adopted as the standard by the Philippine Commission on Higher Education as well as by several state universities in the ASEAN Region affiliated with SEAMEO Regional Centre for Graduate Study and Research in Agriculture (SEARCA) University Consortium.

Development communication is offered at both graduate and undergraduate levels at the University of the Philippine Los Baños (UPLB). Its undergraduate programme consists of 60 units of general education courses, 64 units of development communication courses, 21 units of technical (agriculture, environment, forestry) electives and six units of social science electives totalling 151 units. The product of this programme is thus trained in communication theory and skills, development theory, social sciences and more importantly technical courses adding to their familiarity with sectoral development concerns and contexts.

The graduate programme, however, does not provide for technical electives and general education courses, as expected. It focuses on development theory, communication theory, communication management and research methods.

Multimedia Studies Model. In another campus, the University of the Philippines Open University (UPOU), information and communication studies have been restructured as a singular academic domain deserving of a separate faculty because of the growing interest on ICTs within the professions and the academe. In 2004, the Faculty of Information and Communication Studies became the first academic unit under the UP System as well as the first Open University globally to offer information technology and communication science as integral programmes under one collegiate or departmental roof (since 2004, two other institutions in Southeast Asia, both based in Singapore, established faculties offering information and communication studies as an integral science: the WKWSCI at NTU and the School of Computing at the National University of Singapore (NUS). At the undergraduate level, the faculty offers the Bachelor of Arts in Multimedia Studies (BAMS). The programme is composed of sixteen general education courses (48 units), eighteen major courses (54 units), six residential communication production courses (18 units), five electives (15 units), a three-unit special topics course, and a six-unit special project for a total of 144 academic units.

Graduates are expected to be: knowledgeable with the range and use of multimedia information and communication technologies; articulate in philosophical and theoretical underpinnings of developments in the field and their social implications; abreast with emerging trends, protocols, procedures, and their implications on practice; proficient in hardware operation, software development, and applications use; able to produce multimedia knowledge products; able to contribute to the body of research on multimedia theories and processes; and able to contribute to local multimedia initiatives within the context of global realities. There is also a post-graduate counterpart programme for BAMS.

PROPOSED FORMAL ICTD CURRICULUM

Several universities in Asia have adopted the threefold function of research, instruction and extension, the latter referring to the practice or application of new knowledge in the real world. In this system, research results are expected to contribute to an existing body of knowledge that is then handed over to succeeding generations through instruction, which then paves the way for practice. In this spirit, ICTD research should be seen as part of a continuum that includes instruction and practice. ICTD research results, however, should be situated within a curriculum. What should this curriculum entail?

Curriculum Content. At the undergraduate level, a proposed ICTD curriculum should include: general education (GE) courses; core ICTD courses (computer science, information systems, communication theory, development studies, knowledge management); and technical electives (TE) dealing with specific development sectors depending on the student's predisposition. The ratio of GE:ICTD:TE should roughly be 40:40:20 in percentage value.

At the graduate level, a proposed ICTD programme should be composed of courses on: development theory; communication theory; computer science; information systems; and research, in equal proportions.

Other Considerations. The curriculum should aim for a healthy balance between the technical and social courses. In this regard, applications should not only be taught but analysed as to their social impact. The curriculum should be flexible enough to accommodate the constant upgrades and ever changing nature of technology. Hence, the student should be taught to focus on computer science and basic languages (C++) rather than specific applications and higher languages.

ICTD PROFESSIONAL DEVELOPMENT IN ASIA

The demands of professional development in ICTs. The research projects under the SIRCA programme testify to the increasing pace of diffusion of ICTs within development practice. The projects demonstrate innovative applications across multiple sectors, including; education, health, livelihoods, social services, environmental protection and politics. This should come as no surprise to those who live in developed societies and who take for granted the ever-widening impact of ICTs on most aspects of their daily lives. For the professionals involved in the design and delivery of public services however, the adjustment to an ICT-enabled environment is not always easy or straightforward. Questions arise as to the changing roles and relationships that are usually implied by the effective adoption of ICTs. Additionally, the fast pace of change within the technologies themselves as well as our evolving understanding of how to make best use of them sometimes leads to confusion and inertia amongst those whose activities stand to benefit the most by fully embracing the opportunities that technology offers.

As an example, educators have been known to mistrust the role of ICTs which they see as a replacement for their services rather than as a means for enhancing them. The growth of e-learning was therefore restrained as educators struggled to adjust to their new roles as guides, mentors and designers of teaching material, as opposed to the previous role of instructor in which they felt most comfortable. Similar adjustments have been required within most professional fields as ICTs supplant obsolete practices and are used to re-engineer them. In the medical profession, remote diagnostics aided by ICTs potentially brings the services of highly skilled doctors to locations that never before enjoyed them, creating new opportunities for urban hospitals to serve rural areas. Armed only with reliable ICTs and the knowledge to use them, developing country workers provide digital skills to distant developed country employers in a range of outsourced professions. Such opportunities call for a tech-savvy approach by a wide range of development professional disciplines; actually in all of them — if the full promise of ICTD is to be realised.

How the ICT professional bodies address professional development. The programmes of Asian educational institutions are slowly responding to the dynamics of ICT4D within the development professions and preparing its new entrants for the opportunities that ICTs offer. In the meantime, as these graduates enter and progress through their professions, those that have not benefited from such ICT-informed educational programmes have to grapple with the opportunities that contemporary technologies offer. Moreover, as the technologies themselves rapidly change, knowledge of how to use them that has been painstakingly accumulated rapidly becomes obsolete, and a new learning cycle is set in motion.

In the ICT profession, it has been acknowledged that new technologies are introduced continually, and that existing ones become obsolete almost as soon as they appear. Accordingly, the rapid evolution of the discipline has a profound effect on ICT education, affecting both content and pedagogy. Some observers have suggested that the knowledge acquired in a graduate level qualification in ICTs remains relevant for a maximum of six years. The British Computer Society, a professional body that does not allow easy entry, places high importance on the requirement for keeping up to date with new knowledge in computing. Its code of conduct requires members to "develop your professional knowledge, skills and competence on a continuing basis, maintaining awareness of technological developments, procedures, and standards that are relevant to your field."

The need for professional development amongst ICT professionals is obvious, yet as the technology becomes increasingly approachable by nonprofessionals, there is an equivalent need amongst them to also stay current with technology development. The widespread and growing use of ICTs within most aspects of life and the resultant opportunities for innovative applications with the need for the acquisition of new knowledge have resulted in a continuous transformation of our cultural, social, and political environments. We are living in a world that is constantly impacted by rapid developments in the domains of science and ICTs and where existing knowledge quickly becomes outdated and obsolete. Accordingly, individuals in all professions and the institutions within which they operate have a responsibility to adopt formal measures for maintaining the currency of their knowledge.

What topics to train. The key skill taught to ICT professionals is to learn how to learn. Whilst many will be drilled in the practice of computer programming, or in how to maintain a particular operating system or network, as they progress in their profession, such hands-on skills lose their relevance. They need to be able to understand the theoretical concepts behind these technologies so that they can adapt to the deployment of newer ones. Similarly, there is a body of conceptual knowledge that supports the *use* of technologies and which remains relevant regardless of the technical nature of the particular technology itself. Such 'soft' skills relate to the relationship between technology and society, and to how technology contributes to innovation in a given situation. Analytical skills are required to identify the contribution that technology can make, alongside other, human, factors.

The key ICTD knowledge areas that non-ICTD development professionals require. ICTs have been characterised as 'necessary but insufficient' in their application to business processes, and the same applies to their use for development. Whilst technology enables a variety of usually innovative activities, it is nearly always the actions of people that make the key difference. Accordingly, whilst technically skilled people should be relied upon to deploy the right technology in the right way, it is the other aspects of its use that require attention by development professionals.

In addition to technology, human adaptations are required for desirable results to emerge. They stem from changes in long-standing practices within which individuals have become accustomed and feel comfortable, and from changes in the nature of relationships between various actors involved. For example, the use of mobile phones and the internet to deliver market price intelligence to farmers has often been acclaimed to eradicate the role of middlemen who traditionally hold a monopoly on farming information. Yet one of the SIRCA research projects highlighted how such technology can

actually benefit the middlemen more than the farmers, challenging accepted
wisdom and calling for a nuanced understanding of the precise role of the
technology that exceeds the basics of how it works and how to operate it.
Such situations demand analytical and social skills that can be deployed
under a wide range of contexts and in highly varying situations but which
nevertheless adopt a common conceptual approach based on an understanding
of the relationship between technology and society.

Linking ICTD professional development to ICTD practice. Too often, ICT skills
are regarded solely as technical. The social aspects of ICT use are rarely seen
as being their equivalent in significance. Unfortunately, the skills required
to be proficient in both are usually mutually exclusive. Who hasn't heard of
the computer 'nerd'? Ironically, the softer skills are those that development
professionals are trained in most and seem to possess in abundance, yet
they themselves tend to lack confidence in scenarios that are dominated by
technology. For those who are comfortable amongst people but less so amongst
technology, the basic requirement is to be able to speak and understand, up to
a point, the language of the technologists. This requires some understanding
of ICT concepts without the detail of say a programming language or
networking systems, which will enable the two sets of professionals to form
effective teams.

What do ICTD training development professionals need now, and in the future?
As ICTs permeate most aspects of development practice, development
professionals need to be able to communicate effectively with ICT professionals
and they need to be able to bring in that which is also necessary, beyond
the technology, to make ICTs effective. In the field of corporate ICTs, most
of the university business schools teach Information Systems as a topic that
addresses less of how computers *work* but more of how to *use* them. The
professional Systems Analyst is the product of this learning process and it is a
well recognised role in corporations and public bodies that use computers.

Development practice has yet to arrive at the same level of understanding
about what is required to make ICTs effective and what skills and roles to
adopt in order to make this happen. There are still those in the profession
who doubt that ICTs have a role in development, so the outlook for a rapid
change of approach, which was driven in corporate circles by the urgency
of competitive imperatives that development does not have, is rather poor,
notwithstanding the SIRCA programme. The research that is highlighted
in the programme is far more revealing of the adaptations within the social
contexts of the technology than it is about anything to do with the technology

itself. The narratives are mostly concerned with people; what they did, how they reacted, how they benefited, or didn't. The technology almost forms a backdrop to processes that are fundamentally human and the implication is that these are the processes that need to be carefully managed. All the technology has to do is work reliably.

IMPLICATIONS FOR TRAINING OF MAINSTREAMING ICTD

How governments address ICTD policy and practice within their public services. Most governments in Asia are in the process of formulating ICT policies. Some are well advanced and are in reformulation or fine-tuning mode. Others are just starting out with the beginnings of formal policy-making. The most common approach is to anchor ICTs policies in a programme for electronic government, often alongside deregulation of telecommunications and extension of a national infrastructure. E-government programmes normally begin with the computerisation of back-office government databases, followed by the tentative roll-out of citizen services, often via shared access at community telecentres. De-regulation of telecommunications is swiftly followed by rapid take up of mobile services which soon extend into previously underserved rural areas, sometimes under subsidies from a universal service policy. The discourse surrounding these initiatives is overwhelmingly technical because this is the nature of the decisions that have to be made during these early stages of national-level ICT development.

Evolution of ICTD in Asian governance; some examples. As more public services become available via ICTs, the unaffordability of domestic ICT ownership, most notably among the rural poor, generates a move to shared access at community telecentres. Thousands of such centres have emerged across Asia, fuelling the ICT4D movement and exposing the range and depth of opportunities for using ICTs to reduce poverty. In some countries, telecentre programmes have been incorporated into government ICT policy-making. In India, for example, the government has committed to provide internet access to every village in the country. Similar programmes are under way in Malaysia, Vietnam, Sri Lanka, Bangladesh and the Philippines.

Evolution of ICTD in UN, WB, ADB, etc. The UNDP Asia-Pacific Development Information Programme (APDIP) was launched in 1997 and ended on 31 December 2007. It aimed to promote the development and application of ICTs for sustainable human development in the Asia-Pacific region, through three core programme areas: Policy Development and Dialogue, Access, and Content Development and Knowledge Management. The World Bank's *InfoDev* (Information for Development) Programme began in 1995 and it remains operational. *InfoDev* is a technology and innovation-led development finance programme that supports global sharing of information on ICT4D, and helps to reduce duplication of efforts and investments. The Asian Development Bank (ADB) supports projects that build ICT and telecommunications infrastructure or sector-specific projects that use ICT as a component. No single unit is dedicated to the oversight or operation of all its ICT projects, although some are coordinated by the Regional and Sustainable Development Department. IDRC was one of the first development agencies to embrace ICTs as a key means to foster development and alleviate poverty. It has operated several important projects/programmes, amongst which *the Pan Asia Networking* seeks to understand the positive and negative impacts of ICTs on people, culture, the economy, and society, so as to strengthen ICT uses that promote sustainable development on the Asian continent.

How ICTD has been mainstreamed by the major development agencies. In the early years of the 21st century, there was a period of intense interest amongst development agencies in access to and transfer of technology, especially information and communication technologies, which was spurred on by the UN General Assembly Resolution 56/183 (21 December 2001) endorsing the holding of the World Summit on the Information Society (WSIS). Many of the national telecentre programmes, and the pilot projects that preceded them, were implemented with the aid and support of some of these major development agencies. UNDP is working with the government of Bangladesh and the World Bank underwrote the e-Sri Lanka programme. The IDRC has supported a number of seminal telecentre projects that have contributed significantly to national developments in their countries, e.g. in India and Malaysia.

The development agencies are now tending towards the 'mainstreaming' of ICT4D, which seems to mean disbanding their dedicated ICT4D units and incorporating the activity into their substantive sectoral units; agriculture, health, education, environment and so on. Some argue against this trend, suggesting that subsuming the technology into individual development silos means learning about ICTs becomes trapped by mainstreaming. The specialist knowledge that successful ICT4D deployment requires — about design, development, implementation, evaluation, etc. — does not flow across the silos, causing wheels to be continuously and wastefully reinvented. Also, that mainstreaming agencies fail to take into account ICTs' cross-cutting, integrative capabilities; digital technology's ability to address a whole raft of development goals at once.[1]

LINKS BETWEEN RESEARCH AND PROFESSIONAL TRAINING

How to use ICTD research to build professional training courses. Development professionals need a better understanding of the social aspects of ICTs as well as sufficient awareness of the technical concepts that will allow them to communicate effectively with technical staff. The research that the SIRCA programme has conducted offers insights into some of these aspects and the results can be used to build and strengthen frameworks of comprehension from which rules and generalisations can be synthesised and taught. By adapting these to some of the principles of the information systems discipline, and by combining it with some basic technical knowledge, it becomes possible to build a coherent and teachable body of knowledge around ICT4D.

CONCLUSIONS

The SIRCA programme has demonstrated how ICTs can be usefully embedded into programmes of social, political and economic development. It can now be carried forward in a way that synthesises the experiences into teachable units from which development professionals can learn. Some of the lessons of the programme that can inform such an initiative include:

1. The certainty of unexpected but significant outcomes;
2. The importance of tracking the information chain;
3. The need to identify and differentiate between all the stakeholders;
4. How technology is appropriated by its users;
5. The importance of system design alongside technology deployment;
6. The role of local knowledge in regional planning;
7. The use of referent models of development for design and evaluation;
8. How human responses to technology interventions influence outcomes; and
9. Helping users adapt devices to new uses.

Many civil service organisations have training institutions to promote the professional development of their members. These may be organised within universities or they may be separate from them. They may be supported by development agencies and they may involve the participation of professional associations representing the various skills and disciplines that governments employ. A number of factors contribute to the need for greater engagement of the training that these bodies deliver with the use of ICTs in national development:

1. The increasing use of ICT4D;
2. Diminishing direct involvement by the development agencies;
3. The growth of e-government as the principal delivery platform for public services;
4. The widespread use of mobile phones in rural and lower income communities;
5. The spread of the internet and trend towards broadband;
6. The continuing spread of telecentres; and
7. The emergence of new and lower cost technologies

These factors call for a need for increased emphasis on professional development for ICT4D if its potential is to be fully realised. The research under SIRCA, and other similar studies, reveals the contextual complexities that implementers face and which extend far beyond the technical aspects of the deployment. Unless development professionals come fully to grips with them, opportunities will be lost and ICTs will be regarded as an expensive diversion.

Note

1. Richard Heeks, "Mainstreaming ICTs in Development: The Case Against." In *ICT4D Blog*, 30 October 2011. <http://ict4dblog.wordpress.com/2010/10/30/mainstreaming-icts-in-development-the-case-against>.

9

MULTI-STAKEHOLDER PERSPECTIVES INFLUENCING POLICY-RESEARCH-PRACTICE

Arul Chib, Komathi Ale and May-Ann Lim

Development projects that focus on the use of information and communication technologies for development (ICTD) often involve multiple persons, organisations, methods, and locations that require multiple stakeholder partnerships to enable project success and sustainability. There is a need to develop practical yet theoretically-grounded studies that illuminate the realities of field-based ICTD projects that involve multiple stakeholders, who are inevitably involved in the process. Despite a growing body of literature particularly focused on an evaluation of impact on beneficiaries, there is a gap in research on stakeholder perspectives. We need to understand the complex interactions at the management, operations, and beneficiary levels that occur in the realms of practice, research, and policy. The objective of this study is to define and examine the perspectives of stakeholders in order to develop knowledge on the management of multi-stakeholder ICTD project partnerships. To do so, we develop and propose the *Stakeholder Communication Model*, which begins this process of understanding by focusing on communication patterns between and among various stakeholders in ICTD projects.

INTRODUCTION

The spotlight on ICTs for development emerged in the last decades of the 20th century when the Internet and the Millennium Development Goals (MDGs) were developed. The confluence of these two phenomena resulted in "a new tool in search of a purpose" (i.e. the Internet and ICTs), and the MDGs "were new targets in search of a delivery mechanism" (Heeks 2008, p. 27). A sudden bout of projects, programmes and publications, especially in less-developed countries, occurred, and new, larger stakeholders, such as non-governmental organisations (NGOs) and international development organisations, entered the fray of development work as these two arenas met.

Unfortunately, the newness of the field led to many problems. Early assessments of the field have criticised it as being Western-centric and economically exploitative of the Southern states (Ojo 2004). Pressed to show results, many projects implemented during this period were too technologically deterministic in their outlook; too optimistic in their view of ICTs as a panacea for all development woes (Benner 2004; Ojo 2004; Leye 2007; Raiti 2007; Wieigel and Waldburger 2004). Many case studies have shown that ICTD projects lacked sustainability and scalability, and were mostly anecdotal in nature, lacking critical monitoring, impact and evaluation assessments (Heeks 2008). In recent years, this state of affairs has improved somewhat, with nascent impact assessment methodologies emerging. However, as ICTD interventions have been overeager to adopt new technologies, we see many researchers and practitioners disheartened and cynical when technology does not solve development issues (Alzouma 2005; Kuriyan, Ray and Toyoma 2008; Mansell 1999).

The field of ICTD is also beginning to develop some self-reflexivity as it gains historical traction, although there remain many critical concerns involving ICTD research. Of particular concern are critiques on the lack of coherence, the paucity of practical, yet academically rigorous research, and the field's deficiency in stakeholder research, particularly when dealing with multi-stakeholder projects.

Lack of Coherence in ICTD Research

An important observation of the current ICTD research is the noticeable lack of coherence in the field. An early study by the working group of the United Nations Commission on Science and Technology for Development (UNCSTD) acknowledged the fact that a general 'best practice' ICT model could not be developed due to the disparities between countries (Mansell 1999).

Additionally, ICTD research has been largely anecdotal, and has developed by a diverse set of topics and sectors (Chib 2009). This topical diversity has come at the expense of disciplinary coherence, and has been compounded by a lack of published best research practices that other practitioners can read and learn from (Creswell and Plano-Clark 2007; PAN-IDRC 2007). In order that the field can grow and mature, there is a great need to organise the research processes involved (PAN-IDRC 2007).

Paucity of Practical, yet Academically Rigorous Research

In addition to this lack of topical coherence, there is a paucity of best practices information available on methods and frameworks — a problem that has been noted by the researchers at the PAN-IDR Workshop in Manila 2007 (Creswell and Plano-Clark 2007; PAN-IDRC 2007). They noted that much of the available literature today remains atheoretical and descriptive, and concluded that "too many presentations remained poor, with a lack of theoretical and empirical underpinnings" (PAN-IDRC 2007, p. 2).

Focus on Beneficiaries and Project Processes

The majority of current ICTD evaluations tend to look at the impact that a project has had on beneficiaries in order to assess the project's usefulness, and thus its longevity or sustainability. Many projects are often evaluated against this benchmark, but beneficiaries form only one group of project stakeholders. Whilst it is important that ICTD projects evaluate projects in terms of their beneficiaries and the methods of implementing the project, there exists a paucity of research on the role of stakeholders, such as donors, research teams, and governments, who remain a significantly understudied population in multi-stakeholder projects. Hence, it becomes important to understand the conditions that create conducive environments for ICTD partnerships and collaborations to take place.

Deficit of Multiple-Stakeholder Research

As stated, there is a lack of research being done on projects that involve multiple stakeholders. Stakeholder partnerships are important to ICTD projects, and form the foundation upon which many projects depend on for their success, continuity, and expansion. Even when there have been some attempts at developing multi-stakeholder perspectives, the literature falls short of providing a coherent manner in which to view these interactions.

One reason for the lack of research into stakeholder perspectives in ICTD projects could be due to the fact that, from a methodological perspective, such research and projects are very complex and multi-disciplinary in nature (Raiti 2007). Multiple methodologies often need to be utilised when conducting and analysing the research data, increasing the difficulty of research, especially when it involves multiple disciplines.

One of the reasons behind the lack of effective partnerships is due to the nature of ICTD research as a particular instance of applied research. The impetus behind such research is one that pushes the boundaries of what basic research can illuminate, and is thus more complex in design and implementation (Bhatnagar 2000; Miller 1991). ICTD projects move pure research beyond "basic research", which "focuses on academic issues with no direct concern for the applicability of the results", into the field of "applied research", which is concerned with producing "results that may have practical significance" (Nation 1997, p. 25).

ICTD research is also part of the school of action research that is designed to improve the researched subjects' capacities to solve problems, develop skills (including professional skills), increase their chances of self-determination, and to have more influence on the functioning and decision-making processes of organisations and institutions from the context in which they act (Boog, Keune and Tromp 2003). This process is one that is participatory in nature, and naturally involves a multitude of stakeholders (Chetley 2006; Spinuzzi 2005).

There is a need to bridge the gap between policy and implementation, or practice. Policy recommendations that depend heavily on theoretical models often lack practical considerations, which could lead to implementation problems that stymie the success of ICTD projects. Past success stories of ICTD partnerships and collaboration often involve multiple stakeholders (Batchelor et al. 2003; World Bank 2010). We note that it is critical for the discipline to develop more comprehensively in this direction, especially due to the rising number of researchers conducting ICTD research in the field. As mentioned, many agencies are looking to ICT as a method of aiding the growth and development of the less developed nations. A key point is that a project's success and sustainability could be positively impacted if synergies between different stakeholders and sectors existed.

PRACTICE-RESEARCH-POLICY REALMS

Stakeholders are an important, yet understudied, constituent of ICTD research. Indeed, multi-stakeholder partnerships have been identified as an example of catalyst for change through ICTD projects (Wieigel and Waldburger 2004). However, effective partnerships are rare, and thus understanding the perspectives behind an ICTD project's stakeholders becomes important.

Stakeholders can be categorised by the distinct roles they assume, their objectives, and the ways they influence the project, into three realms — Practice, Research, and Policy. Depending on the realm in which they belong, stakeholders enter at different points of the project with different mandates and interests, bringing with them different goals, methodologies, implementation strategies, reach, and resources, amongst other factors. Crucial to the success of a project's implementation, reach, and impact is the management of these different stakeholder groups. Therefore, the lack of attention being paid to this important component of field research planning needs to be addressed.

It has also been noted that the participatory nature of ICTD action research projects at time creates tensions between stakeholder objectives and deliverables, particularly between the needs of practitioners and researchers (Badham and Sense 2006). However, if a nuanced approach is taken in managing and overcoming the challenges involved in the interaction between sectors, the benefits to development work could be tremendous (Chandrasekhar and Ghosh 2001).

Understanding stakeholder involvement in ICTD projects is important since communication between different groups of project partners is a key ingredient for meaningful interventions (Minore et al. 2004; Reutter et al. 2005). Key players in the ICTD field can be identified through the scope of their projects and funding, as well as the studies and literature that they develop for consumption (Chetley 2006). These players include the agencies traditionally associated with development work, such as *international/ multilateral aid agencies* like the United Nations (UN), the World Bank (WB), the World Health Organisation (WHO), as well as their affiliate agencies like the WB's *infoDev*, the UN Development Programme (UNDP), United Nations Educational, Scientific and Cultural Organisation (UNESCO), the International Telecommunication Union (ITU), and the United Nations Information and Communication Technologies Task Force.

Closely related to the multilateral agencies are the *regional/national/bilateral aid and development agencies,* such as the UK Department for International Development (DFID), Canada's International Development Research Centre (IDRC), Canadian International Development Agency (CIDA), Danish International Development Agency (DANIDA), Swiss Agency for Development and Cooperation (SDC), Swedish International Development Cooperation Agency (SIDA), and others. These groups in the policy realm primarily derive their funding from their respective governments. In the research realm, *academic institutions* have also developed expertise in this area, and have contributed greatly to this field. Another important group of stakeholders are the *non-governmental organisations (NGOs)* in the practice realm. Many ICTD projects involve almost all of these stakeholders in the planning, implementation, evaluation, and continuity of their work.

The existing literature does not, however, include practical, applied frameworks on how to conceptualise, format and implement a multiple stakeholder project (Creswell and Plano-Clark 2007). Though theories have not directly addressed the examination of stakeholders, a review of extant models suggests key learning points. Concerned with impact assessments of ICT intervention projects, Heeks and Molla (2009) built on the standard input-process-output model, and created the ICTD Value Chain model. It divides assessments into four targets: readiness, availability, uptake, and impact. Although framed as an evaluative model, Heeks and Molla's model provides a perspective of the different entry points that stakeholders may engage in ICTD interventions. Additionally, this model adds another dimension for ICTD project implementers' consideration — time. It suggests that over time, stakeholder foci may shift and change, which is another important, yet nebulous facet of ICTD project implementation. The model offers the perspective that there are important precursors that must be in place before, during, and after the ICTD project takes place — for example, institutional or organisational systems, data systems, human resource, and technical availability. The commitment of these resources to the project by partners and other stakeholders is important to the project. However, there are limitations to this model — despite being fairly comprehensive in terms of over-viewing the processes involved in ICT projects, it does not specifically discuss stakeholder contribution in project success or failure. This approach misses out nuances of communication between people and organisations involved, often happening before, during, or in-between the processes detailed.

The extended Technology-Community-Management (TCM) model (Chib and Ale 2009; Chib and Zhao 2009; Lee and Chib 2008) serves as a useful prism to view ICTD projects and stakeholder involvement. The TCM model proposes that three factors of technology, management, and community, will lead to sustainable ICT interventions. Whilst technology focuses on software and hardware components, project management entails financial and regulatory planning, and developing key partnerships. Community involvement, by understanding needs, sharing ownership, and providing technology training, is an important element in achieving sustainable impact. Further, the model focuses on the barriers to development, but fails to take into account organisational barriers related to stakeholders involved with ICTD projects, particularly at various key phases of project implementation. Further, focusing on the beneficiary community, the model fails to detail how stakeholders interact and communicate in terms of their internal and external structures, such as the decision-making hierarchy, opinion leadership, and information and communicative links.

Stakeholder Communication Model

The Stakeholder Communication Model proposes that there are several groups of stakeholders involved with any ICTD project, operating on the Policy-Research-Practice realms at different levels: management (top), operations (middle), and beneficiary (bottom) levels. (We recognise the implicit hierarchy in such a categorisation, but continue for the sake of ease of understanding. We trust that the model is read in the spirit in which it is written — that bottom-up emergent practices are as important as top-down hierarchies.) The *management* level refers to decision-makers in the stakeholder organisations, whilst *operations* refers to stakeholders involved in carrying out the project tasks, and *beneficiaries* refers to the people who are either directly benefiting or are interacting and communicating with the beneficiaries of the project. The iterative interactions that occur in the realms of practice, research, and policy should be considered; this is essential in forging the continuity of projects. These levels of stakeholders are depicted in the Stakeholder Communication Model.

With the Stakeholder Communication Model, stakeholder communication pathways can be mapped in two directions: horizontally and diagonally between adjacent stakeholder groups on the inter-stakeholder level, and vertically between layers of intra-stakeholder groups. This can happen within stakeholder, or from one level of stakeholder to another.

MULTI-LEVELS OF STAKEHOLDER COMMUNICATION

The Stakeholder Communication Model provides a conceptual lens by which to define and explore communication pathways between and amongst stakeholders involved in ICTD processes. It offers a new stakeholder perspective that the previous models lacked by proposing that ICTD stakeholders fall into three stakeholder levels: management, operations, and beneficiaries, within the Practice, Research, and Policy realms. This model enables the systematic exploration of the conditions within which ICTD projects may be implemented, especially in terms of the management of stakeholders.

FIGURE 9.1
Stakeholder Communication Model

	PRACTICE	RESEARCH	POLICY
MANAGEMENT	Project Management Team	Principal Investigators	Funding Institutions
			Community Leaders
			Government Bodies
OPERATIONS	Technology Developers	Field Researchers	Local Administrators
	Project Execution Team		
BENEFICIARY	Project Beneficiaries		

INTER-STAKEHOLDERS   ↑ INTRA-STAKEHOLDERS
←——→ ⤬   ↓

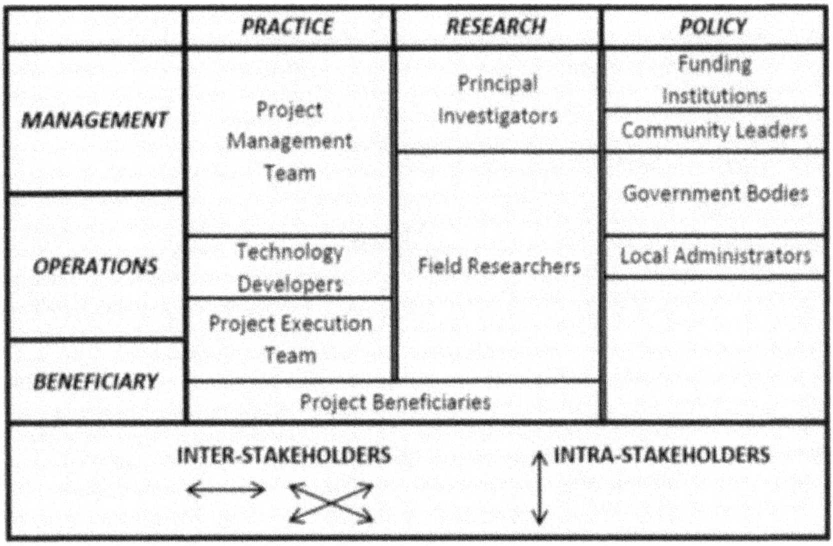

Source: Authors.

The development of this model adds a new dimension of multi-stakeholder communication to the available literature on ICTD projects that has been critiqued as lacking academic rigour (Leye 2007). The earlier critique of existing models is remedied with this new model, especially through its observation that communication pathways occur along three levels of stakeholders. The model helps practitioners understand the organisational challenges involved when working and communicating within a multi-stakeholder project. The Stakeholder Communication Model offers a scaffold by which we can frame these stakeholder communication pathways, especially by detailing the internal hierarchy of stakeholders.

Management-Level Stakeholders. We identify several crucial issues for stakeholders functioning at the management-level. First, it is critical that stakeholders approach ICTD projects with an understanding of the involvement of different levels of stakeholders — from management, to operations, to the beneficiaries or sub-contractors who work on this level. The initial project scope, in terms of organisational hierarchy, directly impacts the expectations and deliverables from each stakeholder involved.

Initial communications preceding the formalisation of the ICTD project should also include goal alignment for all stakeholders involved, ensuring that stakeholders on this level understand other stakeholders' objectives and goals. Understanding partners' expectations will help shape formal participation agreements, such as project agreements or terms of reference, which are important documents needed for the smooth running of ICTD projects. These documents may eventually function as communication bridges for the operations-level stakeholders.

Operations-Level Stakeholders. Operations-level stakeholders should seek to constantly renew the communications pathway between their work and management-level stakeholders, in order that institutional needs and implementation methods stay synchronised with each other. This is important especially when there is a high turnover rate of staff on both levels. A systematised method of communication would help project continuity in light of these personnel changes. Although most staff on the operations level function well across organisations, staff within organisations should work together to ensure the smooth transition of roles and responsibilities, especially as individuals transit in and out of the project.

Operations-level stakeholders should also be cognisant of the levels of organisational hierarchy that exist on the beneficiary level. This is a management-level challenge, and applicable on a smaller scale to the operations level stakeholders. They should work to communicate the purposes of the project on all levels, so as to ensure that all beneficiary-level stakeholders understand and participate in the ICTD project to the best of their ability.

Beneficiary-Level Stakeholders. The communications within, and between, beneficiary-level stakeholders can be hampered by the unique challenges of working in the field. Infrastructural and technological deficits in physical and human resources create issues for data-collection between people, as well as for automated data-collection. Beneficiary-level stakeholders would do well to use these challenges as opportunities to build their capacity for solving problems.

CONCLUSION

As there was a clear need to add stakeholder perspectives to the field of ICTD research, especially when exploring multiple-stakeholder projects, the

Stakeholders Communication Model contributes to the literature by offering a guiding framework to view stakeholder communications. It found that various stakeholders function on multiple levels — management, operations and beneficiary levels. By using the Stakeholder Communications Model to map each stakeholder level's communication pathways with each other, it offers a systematic view for ICTD project implementers to view and manage different levels of communication that occur in such projects. It is hoped that future implementers of ICTD projects will formulate better, more sustainable projects through the nuances of inter- and intra- stakeholder communication pathways viewed through the conceptual lens of the Stakeholder Communication Model.

References

Alzouma, Gado. "Myths of digital technology in Africa." *Global Media and Communication* 1 (2005): 339–56.

Badham, Richard J. and Andrew J. Sense. "Spiralling up or spinning out: A guide for reflecting on action research practice." *International Journal of Social Research Methodology* 9 (2006): 367–77.

Batchelor, Simon, Soc Evangelista, Simon Hearn, Malcolm Peirce, Susan Sugden and Mike Webb. *ICT for development: Contributing to the millennium development goals-Lessons learned from seventeen infoDev projects*. Washington DC, USA: World Bank, 2003.

Benner, Caroline. "De-Hyping IT." In *Foreign Policy*, 1 September 2004. <http://www.foreignpolicy.com/articles/2004/09/01/de_hyping_it> (accessed 10 January 2011).

Bhatnagar, Subhash. "Social implications of information and communication technology in developing countries: Lessons from Asian success stories." *The Electronic Journal of Information Systems in Developing Countries* 1 (2000): 1–9.

Boog, Ben, Lou Keune, and Coyan Tromp. "Action research and emancipation." *Journal of Community and Applied Social Psychology* 13 (2003): 419-425.

Chandrasekhar, C.P. and Jayati Ghosh. "Information and communication technologies and health in low income countries: the potential and the constraints." *Bulletin of the World Health Organisation* 79 (2001): 850–55.

Chetley, Andrew. "Improving health, connecting people: the role of ICTs in the health sectors of developing countries: A Framework Paper." In *infoDev*, May 2006. <www.infodev.org/en/Publication.84.html> (10 January 2011).

Chib, Arul. "The role of ICA in nurturing the field of information and communication technologies for development." *International Communication Association newsletter* 37 (2009): 13–16.

Chib, Arul and Komathi Ale. "Extending the Technology-Community-Management model to disaster recovery: Assessing vulnerability in rural Asia." *Proceedings of the 3rd International IEEE/ACM Conference on Information and Communications Technologies and Development (ICTD)*, 17–19 April 2009. Doha, Qatar: Carnegie Mellon.

Chib, Arul and Jinqui Zhao. "Sustainability of ICT interventions: Lessons from rural projects in China and India." In *Communicating for social impact: Engaging communication theory, research, and pedagogy*, edited by Lynn Harter and Mohan J. Dutta. ICA 2008 Conference Theme Book: Hampton Press, 2009.

Creswell, John W. and Vicky L. Plano Clark. *Designing and conducting mixed methods research.* Thousand Oaks, California: SAGE Publications, 2007.

Heeks, Richard. "ICT4D 2.0: The next phase of applying ICT for international development." *Computer 41* (2008): 26–33.

Heeks, Richard, and Alemayehu Molla. *Impact assessment of ICT-for-Development projects: A compendium of approaches.* Manchester UK: Development Informatic Group, Institute for Development Policy and Management, 2009.

Kuriyan, Renee, Isha Ray and Kentaro Toyama. "Information and communication technologies for development: The bottom of the pyramid model in practice." *The Information Society* 24 (2008): 93–104.

Lee, Seungyoon and Arul Chib. "Wireless initiatives for connecting rural areas: Developing a framework." In *Participation and media production: Critical reflections on content creation* — The ICA 2007 Conference Theme Book, edited by Nico Carpentier and Benjamin De Cleen. Newcastle, UK: Cambridge Scholars Publishing, 2008.

Leye, Veva. "UNESCO, ICT corporations and the passion of ICT for development: Modernization resurrected." *Media, Culture and Society* 29 (2007): 972–93.

Mansell, Robin. "Information and communication technologies for development: Assessing the potential and the risks." *Telecommunications Policy* 23 (1999): 35–50.

Miller, D. C. *Handbook of Research Design and Social Measurement.* 5th ed. California: Sage Publications, 1991.

Minore, Bruce, Margaret Boone, May Katt, Peggy Kinch and Stephen Birch. "Addressing the realties of health care in northern aboriginal communities through participatory action research." *Journal of Interprofessional Care* 18 (2004): 360–68.

Nation, Jack R. *Research methods.* Upper Saddle River, N.J.: Prentice Hall, 1997.

Ojo, Tokunbo. "Old paradigm and ICT4D agenda in Africa: Modernization as context." *Journal of Information Technology Impact* 4 (2004): 139–50.

PAN-IDRC. *PAN-IDRC Conference on the Effects of the Information Society and Workshop on Research Methodologies.* Manila, Philippines, 2007.

Raiti, Gerard C. "The lost sheep of ICT4D research." *Information Technologies and International Development* 3 (2007): 1–7.

Reutter, Linda, Miriam J. Stewart, Kim Raine, Deanna L. Williamson, Nicole Letourneau and Sharon McFall. "Partnerships and participation in conducting poverty-related health research." *Primary Health Care Research and Development* 6 (2005): 356–66.

Spinuzzi, Clay. "The methodology of participatory design." *Technical Communication* 52 (2005): 163–74.

World Bank. *Poverty*, 2010. <http://data.worldbank.org/topic/poverty> (accessed 2 November 2010).

Weigel, Gerolf, and Daniele Waldburger. *ICT4D — Connecting people for a better world. Lessons, innovation and perspectives of information and communication technologies in development.* Berne, Switzerland: Swiss Agency for Development and Cooperation and the Global Knowledge Partnership, 2004.

10

FROM PRODUCTION... TO DISSEMINATION... TO ADOPTION

Ma. Regina M. Hechanova

As researchers, we seek to resolve unanswered questions; test answers presented by others and refine our understanding of a given phenomenon. After developing our proposals, gathering and analysing data, we write and submit our articles to scholarly journals hoping for a nod from our peers. When we do get such affirmation, we give ourselves a pat on the back, happy that we have done our share in building knowledge. But have we really? When can we say that the research we do makes a difference? Journals today report impact factor — a number that reflects how many other scholars have cited one's work. Although a decent indicator, has it been widely accepted merely because it is convenient? At the end of the day, shouldn't impact be ultimately assessed in terms of whether our work has reached the people who shape policy and decisions and have actually led to positive change?

Of course, such a question presupposes that government and industry leaders, policy-makers and practitioners actually read academic publications. Yet a criticism sometimes hurled at academics is that our research is isolated from the real world. And our critics are not entirely wrong. Some academics formulate research problems based on a review of literature without input from current realities. They get caught up in dissecting a phenomenon — even when the world has since changed and moved on.

RESEARCH THAT MATTERS

It is not the answer that enlightens, but the question. — Decouvertes

Research as a Knowledge Product

One way to ensure that the research we do matters would be to take some lessons from business management. What if we viewed research as a knowledge product? A basic principle in strategic management is 'know your market'. Business people will naturally care about providing products and services that are marketable. Applying this principle as researchers, we need to ask ourselves 'who is our audience?' If our study will only benefit a small population or a specific organisation, its chances of getting published is unfortunately, nil. If we wish to have a greater market, we need to ask questions that will ultimately impact not just a single organisation or community. In the same vein, we need to decide — are we targeting a global or local audience? If we are asking questions that only people from one locale would appreciate, then that limits us to local outlets. But if we are able to frame our study to apply to say, other developing countries, or other Asian countries, then there is a greater chance of it being interesting to international audiences.

Market research seeks to understand customer behaviour. What do they care about? What do they do? What are their needs? Similarly, we need to ask ourselves what are the hot topics that researchers, practitioners or policy-makers care about? What are the controversial topics that people debate about? What are the questions those in the field constantly ask that we have no answers for? The answers to these questions will be evident when we keep abreast of current events, attend conferences, or engage in dialogue with colleagues, policy-makers and practitioners.

The salient questions are likely to be different depending on whom we ask. Those involved in academic research are mostly concerned with extending or developing theory or methodology. Journal editors, reviewers and grant agencies will ask — 'what is new about this research that has not been reported?', 'how does this build on current theory?' or 'what new methodology or lens is being used in investigating this topic?' Practitioners in industry and non-governmental organisations, on the other hand, are not as concerned about theory but the application of knowledge to address current and future challenges. They are likely to ask 'how can this research help me solve my challenges or goals?' and 'how do I use the knowledge to improve how we do things?' Finally, national and international policy-makers and government leaders will be concerned about information that will shape policy and laws. Thus their interest will be in finding solutions to social problems by providing safeguards or establishing an environment that will address social needs. Their questions may include, 'how serious is this social problem and what are its causes?', 'what interventions and regulations need to be put in place?' They may also be concerned about investigating the effectiveness of an intervention or policy.

Research as a Solution

A problem is certainly another good starting point for determining what matters. When there is a felt need, we not only have a potential audience for our findings, we also have potential partners for conducting the research. In the context of applied research, research may be motivated by desired goals such as poverty alleviation, development or economic growth. Beyond goals, however, needs are other starting points for developing relevant research problems. What are the problems people face that need solutions? What information do decision-makers need in order to craft policy or design programmes? Even when there are no immediate problems, knowledge generation can also be positioned as instruments for growth. Are there opportunities that can be maximised given more information?

Research as Development

If research can be used to determine appropriate solutions or interventions, it is just as critical in assessing their effectiveness. The cycle of diagnosis, action and evaluation provides a means for elevating practice and knowledge creation. What are the solutions or interventions that have been put forth and is there adequate evidence of their effectiveness? What is the impact of certain policies that have been implemented?

Basic vs. Applied Research

I am certainly not advocating purely applied research. I am sure there are topics that practitioners care little for but academics think are important. Nobody believed that the world was round until Ferdinand Magellan navigated it. But he did it anyway, and later proved that he was right. The reality is that problems that may interest academics may be different from those that interest practitioners. However, basic and applied research paradigms are not necessarily incompatible. The key is to find that intersection between what people care about and what has scientific merit. At the Ateneo Center for

Organisation Research and Development (Ateneo CORD), we have found some success with this. In setting our research agenda, we begin by asking practitioners about their goals and issues. Their inputs are then further refined by a review of the literature. When asked to conduct applied research, we may piggyback some questions for more basic research problems when permitted by clients. Conversely, we sometimes analyse results of basic research in different ways — using more descriptive analysis for practitioners and using more sophisticated analysis for other academic audiences.

FROM KNOWLEDGE PRODUCTION
TO DISSEMINATION

If a tree falls in a forest and no one is around to hear it, does it make a sound? — Mann & Twiss

Our goal as researchers is to be heard. For many academics, being heard typically means being read by other scholars. Unfortunately, academic publishing has become increasingly difficult. Competition amongst academic institutions has reinforced this and put pressure on academics to publish in "first-tier" journals. Such demand has created a cut-throat environment and many of these journals only publish one in every ten submissions.

The peer-review process has also been criticised for its length (Benos et al. 2007). A journal article will pass through a main editor first before it is fielded to peer-reviewers. If one is lucky, one is asked to revise and re-submit. This process can take anywhere from one to two years, and sometimes even more. Such a lengthy turn-around time is especially problematic in dynamic fields such as ICT because information may be obsolete by the time it is published.

We look to our colleagues to obtain validation that our work has worth. Although the process of peer review is a means to assure quality, it is still a fairly subjective process and can be biased. Reviewers and editors have the power to shape a manuscript and unless authors meet referees' demands, they are unlikely to have their manuscript published. A recent study amongst researchers show that a fourth of authors report that they were obliged to make changes to their manuscript that they did not agree with (Bedeian 2003).

But perhaps most relevant to the issue of dissemination is the problem of accessibility. Knowledge production has become a business and given the current publication model, most top-tier journals are only available via paid-access. This means they are inaccessible not just to the general public, but also to researchers in developing countries whose universities cannot afford subscription to these journals. Given the limited accessibility of these traditional outlets, it is important that we explore other avenues if we wish for our work to reach a greater audience.

Open-Access Journals

Open-access (OA) journals present a solution to the problem of dissemination and accessibility. Advocates of OA journals claim that they enable research articles to be published more quickly and widely compared to traditional print journals. Ironically, even as the number of OA journals and publications has increased in the past decade, a study amongst scholars reveals that only 30% intend to publish in these journals. Scholars are afraid that publishing in these journals would negatively affect their promotion and tenure. This fear is motivated by the perception that OA journals have low impact factor and receive low citation rates (Xia 2010). However, the jury is still out on this — some studies provide evidence of greater citations in OA journals (Eysenbach 2006), whilst others refute its advantage (Umstattd et al. 2008). But salient to the issue of accessibility is a study that reveals a growing interest amongst policymakers on the use of online and especially open access research (Willinsky 2004).

Conferences

A common avenue for dissemination is conferences. Whether professional or thematic, presenting research findings to peers and stakeholders provide a means for information to be readily heard. At the same time, researchers obtain input that may enrich their insights and conclusions.

SIRCA, for example, provided an opportunity for its researchers to present their findings at the ICA 2010 pre-conference session, Innovations in Mobile Use 2010 in Singapore and the Strengthening ICTD Research Capacity in Asia (SIRCA) Final Conference in April 2011. Individual researchers in SIRCA also presented at various other conferences. For example, Pham Huu Ty (2011) presented his findings on the use of GIS in erosion and landslide hazard mapping to colleagues at Hue University of Agriculture and Forestry (HUAF), Vietnam and at the Utrecht University in the Netherlands. According to SIRCA mentor Richard Heeks (2011), there is value to these dissemination efforts, "These tend not to appear on academic citation rankings, but have been particularly useful in building what one might call 'academic capital' — the knowledge of one's ideas; reputation; and contacts that have a long-term value."

Dissemination Forums

Beyond academic conferences, other researchers organise forums to disseminate their findings. Compared to conferences, forums tend to be smaller and more homogenous in terms of topics and participants. Grace Mirandilla Santos (2011) presented the results of her research on 'The Filipino Blogosphere and Political Participation in the Philippines" by organising her own forums in Manila, Cebu, and Davao. Working with local blogger societies, the forums featured her research as well as other invited speakers who discussed various aspects of political blogging. "I invited bloggers, readers, students, members of the academe, NGOs, and media to attend these events. The forums were able to get some media and top blog exposure, so that translated to a wider audience," says Mirandilla-Santos.

Community Feedback

When research is being done in the context of development — the critical stakeholders are the practitioners, decision-makers and organisations/ communities who can benefit from the knowledge. A common intervention used in organisation and community development is survey feedback. In this process, data-gathered is fed back to the organisation/community and action-planning is done based on the results. Pham Huu Ty, and his team, disseminated the results of their research on the use of GIS in erosion and landslide hazard mapping directly to the communities they gathered data from. His mentor, Richard Heeks (2011), says:

> We were fortunate that this was an action research project, so our view that the 'research audience' encompasses communities and practitioners was readily encompassed. Integral to the research was a process of feeding back findings to community members and local officials. This helps ensure that knowledge does not just remain academic, but leads to development-related actions. I wouldn't want to over-play it, but I think there is also an ethical dimension here that ICT4D researchers can consider — we expect our research subjects in poor communities to give us their time and their data, but what do we give them in return?

Making Effective Presentations

Other than finding the right venues to disseminate our results, what is likewise critical is being able to present our findings effectively. Yet academics sometimes take it for granted that speaking in front of laymen is different from sharing our findings with peers. The worst thing we can do is to deliver a jargon-filled lecture where people come away scratching their heads wondering what we had just said. When training younger researchers on making effective presentations, here are some tips I give:

- *Know your audience.* Put yourself in their shoes and remember what language would be most appropriate. Use their words and avoid jargon they would not understand.
- *Begin with a bang.* Begin with something that would interest them — an experience they can relate to, a startling fact that would make them sit up, a question that would make them think, a picture that would catch their eye.
- *Keep it short.* Today's presentations are getting shorter and shorter and a 15-minute presentation is often the norm rather than exception. Determine what the most important content is — often that is just your research problem, results and implications.
- *Use visuals.* We remember more what we both see and hear, so visuals are important. One rule of thumb is that you should *not* have more slides than minutes. Use key phrases instead of sentences and graphics in lieu of words.
- *Maintain eye contact.* Be sure to look your audience in the eye. Don't read from notes or the project screen because you will lose eye contact. Don't turn your back to the audience or stand between the project and the screen.
- *Practice, practice, practice.* The only way to have a smooth presentation is to prepare and practice. Ask someone to observe you and give you feedback — Do you use a lot of non-words like 'uh'? Do you have distracting mannerisms? Do you move too much? Are there things that you should add or omit from your presentation?

Popular Media

Because opportunities for face-to-face presentations can be quite limited, researchers may wish to explore using popular media to disseminate their work. As part of SIRCA, Dr Ang Peng Hwa and I examined the viability of online counselling for migrant workers. The project involved setting up a portal that would provide free online counselling for overseas Filipino workers care of counsellors from the Psychology Department of the Ateneo de Manila University. I shared our results on television news and talk shows. I also wrote an article about the project that was published in the country's largest newspaper (see Figure 10.1). SIRCA set up a website[1] for its investigators to share information and photos.

FIGURE 10.1
Newspaper Column

Source:

Digital Media

Mentor, John Traxler (2011) also encourages researchers to explore joining and contributing to online communities in every format — *tweets, blogs, wikis, and lists.* SIRCA researcher, Grace Mirandilla-Santos did this through her blog,[2] *Facebook*, and *Twitter* accounts to inform friends and colleagues about her study and the forums. The forum was also featured in online articles and blogs (Figure 10.2).

Writing for Laymen

A couple of years ago, the Ateneo Center for Organisation Research and Development partnered with the Philippines' largest newspaper for a bi-monthly research-based column. Essentially, we report results of research but in a way that the general public would understand. Initially, the column was quite a challenge because the style and length requirements in media writing are so different from scholarly writing. Unlike a typical academic article where you may be given 30,000 words, media pieces are limited to 500 to 1,000 words.

I have since learned to utilise principles in creative writing to make my media articles more readable. For example, a critical part of any article is its opening. It's important for us to grab the reader's attention and to do this we can make use of a number of techniques: an anecdote, controversial information, a quote, a description, a question, a direct address or a preview of the article. Unlike writing technical articles that tend to use complex sentences and a lot of jargon, I keep my words simple and sentences concise. I make my paragraphs short and use subheadings to make it easier for my reader to follow.

FROM DISSEMINATION TO ADVOCACY

Knowing is not enough; we must apply. — *Leonardo da Vinci*

Building knowledge is one thing. Ensuring that knowledge gets applied is another. Unfortunately, the path from research to practice is a long one and requires more than dissemination. Policy, structural and systemic changes require collaboration and working with other institutions that can help effect change.

Change management theorists emphasise the importance of obtaining the buy-in and support for desired change (e.g. Kotter 2002). If we wish to

FIGURE 10.2
Live Blogging Screen Capture

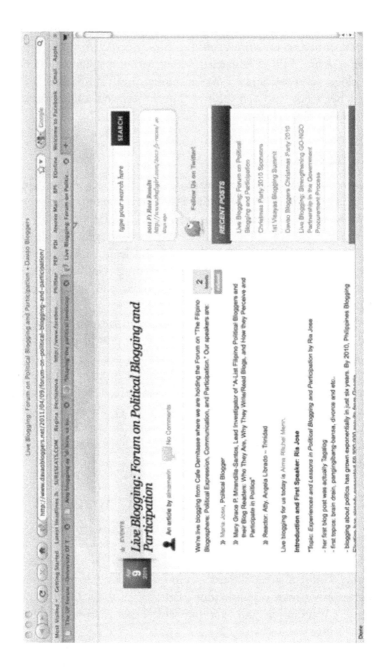

Source:

see the knowledge built adopted, we need to know who are the key decision-makers we need to reach and how to best reach them. Fortunately, there appears to be a growing appreciation for an evidence-based approach to policy and programme formulation. This presents an opportunity for researchers to contribute and highlight the value of research amongst policy-makers.

Our efforts will be in vain, however, if our academic institutions continue to put more emphasis on academic dissemination rather than utilisation. Unfortunately, rewards and promotions in educational institutions are based more and more on academic publications. Although understandable, it does not encourage researchers to pursue other forms of dissemination. I am fortunate that the Ateneo de Manila University has begun encouraging these alternative forms of dissemination. One of its projects for its 150th anniversary was a book intended for leaders and policy makers. Called 'Agenda for Hope' the project brought together a multi-disciplinary team of scholars who were asked to write in layman's language about relevant social issues. In addition, even as it continues to reward publications in first-tier journals, the Ateneo has also created awards for research with the most social impact. Candidates for this award are asked to showcase evidence of impact from communities or institutions. Although this paradigm is still evolving, my hope is that it continues to deepen and serve as an example for others.

> It is science alone that can solve the problems of hunger and poverty, of insanitation and illiteracy, of superstition and deadening of custom and tradition, of vast resources running to waste, or a rich country inhabited by starving poor... Who indeed could afford to ignore science today? At every turn we have to seek its aid... The future belongs to science and those who make friends with science.
>
> — Jawaharlal Nehru

Whilst I heartily agree with Nehru, I cannot help but wonder — if, indeed, science is the solution, why is it that we can lay claim to unprecedented scientific breakthroughs yet still live in a world mired in social problems? Perhaps it is because our concept of the cycle of science is incomplete. If science is to really make a difference, academic institutions can no longer afford to be ivory towers and we who build knowledge need to truly embrace the responsibility of disseminating knowledge to those who need it most.

Notes

1. "Strengthening ICTD Research Capacity in Asia: Current Projects," last modified 2011. <http://sirca1.sirca.org.sg/current-projects/>.

2. "Grace Notes," last modified 22 July 2011. <http://gracemirandilla.tumblr. com/>.

References

Bedeian, A. G. "The manuscript review process: The proper roles of authors, referees, and editors'." *Journal of Management Inquiry* 12 (2003): 331–38.

Benos, Dale J., Edlira Bashari, Jose M. Chaves, Amit Gaggar, Niren Kapoor, Martin LaFrance, Robert Mans, David Mayhew, Sara McGowan, Abigail Polter, Yawar Qadri, Shanta Sarfare, Kevin Schultz, Ryan Splittgerber, Jason Stephenson, Cristy Tower, R. Grace Walton and Alexander Zotov. "The ups and downs of peer review." *Advances in Physiology Education* 31 (2007): 145–52.

Eysenbach, Gunther. "Citation advantage of open access articles." *PLoS Biology* 4 (2006): 692–98.

Heeks, Richard. Personal correspondence, 10 May 2011.

Kotter, John P. and Dan S. Cohen. *The Heart of Change.* Boston, Mass: Harvard Business School Publishing, 2002.

Mirandilla-Santos, Grace. Personal correspondence. 9 May 2011.

Pham Huu Ty. Personal correspondence. 10 May 2011.

Umstattd, Laura J., Marcus A. Banks, Jeffrey I. Ellis and Robert P. Dellavalle. "Open access dermatology publishing: No citation advantages yet." *The Open Dermatology Journal* 2 (2008): 69–72.

Traxler, John. Personal correspondence, 11 May 2011.

Willinsky, John. "As Open Access Is Public Access, Can Journals Help Policymakers Read Research?" *Canadian Journal of Communication* 29 (2004): 381–401.

Xia, Jingfeng. "A longitudinal study of scholars' attitudes toward open-access journal publishing." *Journal of the American Society for Information Science,* 2010.

SECTION III
Research Outputs

11

IT'S THE TALK, NOT THE TECH
What Governments Should Know
About Blogging and Social Media

Mary Grace P. Mirandilla-Santos

ABSTRACT

The emergence of Web 2.0 gave rise to low-cost, user-centric, interactive new media that promote social networking and make almost anything and anyone with internet access within 'clicking reach.' The fast uptake and growing usage of social media — even in countries where internet penetration is relatively low — are an indication of their immense popularity. Their diversified application over the past years, especially in the realm of politics, has been changing the nature of political communication and the landscape in which citizens engage government and public institutions, and in political processes. The undeniable disruptive impact of social media, like blogs, in sharing information, opening communication lines, and gaining more access to both government and citizens, is enough to overwhelm any scepticism that the internet has changed the game and continues to effect irreversible changes in politics and governance. The question is: how can governments put social media into meaningful use? Based on a SIRCA-funded study (Mirandilla-Santos 2011) on the actors behind the Filipino political blogosphere, this thought piece offers some practical insights on how social media, particularly blogs, can be used as a potentially effective governance tool by gaining a critical understanding of the citizens who actually use them. It provides a snapshot of active *netizens*: who they are, why they access political blogs, how they perceive and participate in politics, and whether and how blogs are used for political participation. This article outlines how blogs and other social media are used in facilitating political information and communication, as well as their potential for deliberation and effecting social change. It will also discuss the catalytic effects of converging blogs with social networking sites, like *Facebook*, and micro-blogging site, *Twitter*, and how they can be used in governance. Finally, this article will examine the potential for an emerging governance approach in the Philippines, wherein most government agencies now use various social media. This new development is both a great opportunity and huge challenge for democratic engagement. Whilst still in the nascent stage, this approach has undeniably increased the potential and expectation for transparency. However, the Philippines' long history of closed-door policy-making, bureaucratic processes, and elitist decision-making tempers the promise of social media in transforming governance and politics. A reality check of the maturity of internet usage by Filipinos is also imperative. A study in 2009, for example, revealed that "normalisation" was at play in cyber-campaigning for the May 2010 presidential election, as most campaign websites focused mostly on providing a candidate's personal information, whilst missing opportunities to more actively engage voters. The networking and mobilisation features of websites remained underutilised, although online calls to join a candidate's "team" of core supporters became common (Mirandilla 2009). With the expanding reach of mobile technology and increasing popularity of social media, the marriage of internet and politics is at a critical juncture. How governments and citizens use social media to engage each other today will forever change the landscape of governance.

INTRODUCTION

From the time the first email was sent, there was no doubt that the internet was bound to change many aspects of how we live, learn, make a living, and even love. It is, therefore, no surprise that it has also impinged on the realm of politics and governance. In many ways, the internet, as a networked and open medium, is changing the nature of political communication and the landscape of participation. Some scholars, like Grossman (1995), argue that electronic media can enable citizens to participate in politics whilst others, such as Putnam (2000), are more critical of electronic media, predicting that usage would decrease social capital and inhibit participation. Still, others see it as a double-edged sword (Cooper and Brackman 2009; A. Howard 2011), a powerful tool, or a weapon if you will, whose full capacity has yet to be fathomed by the people who created it.

Whilst the internet allows fast access to unlimited sources of information, the same attribute could also sacrifice accuracy and accountability, unlike the more stringent 'virtues' of journalism (Singer 2005). The internet can help promote interaction among the public and with government, which can increase learning about political issues and processes (Hacker 2000 cited in Tedesco 2007) and help overcome cynicism (Tedesco 2007). However, exposure to political messages online may also serve to either heighten or moderate the level of political cynicism, depending on how the political statement is communicated (Sweetser and Kaid 2008). The internet offers more opportunities for gaining information from multiple viewpoints, which, combined with discussion, can enhance political efficacy (Nisbet and Scheufele 2004 cited in Tedesco 2007; Kenski and Stroud 2006). This same transparency that allows unfiltered views and comments to be posted on a government agency's blogsite or *Facebook* wall also means allowing everyone, regardless of intention or political view, to express his or her opinion. This could include posting content that malign (invectives and personal attacks), distorted facts, cyber-bullying, and spam messages. The other side of this is government using the internet for surveillance and censorship (see e.g., OpenNet Initiative; Kelly and Cook 2011).

The nature and extent of the internet's capacity reside in emotional, often unpredictable, and biased humans, after all. No other internet tool was built more congruent to individual interests, moods, and quirkiness than new media, defined as the shift of production, distribution, and communication of data or information to computer-mediated or digital formats. Compared to 'old' media that are linear, hierarchical, and affect one stage of communication, new media affect the "acquisition, manipulation, storage and distribution of information" and "all types of media — text, still images, moving images, sound, and spatial constructions" (Manovich 2001). And with the advent of Web 2.0, we saw an explosion of low-cost, user-centric, and highly-interactive online applications, such as blogs and social networking sites, which allow any individual to be a content creator and interact with an online social network whose reach is measured in global proportion. Hence, these new media tools are more aptly called 'social media', as they are "targeted at forming and enabling a potentially massive community of participants to productively collaborate" (Bradley 2010) and adopt a "bottom-up, many-to-many" publishing model (Campbell 2010).

Now, one can argue that the impact of new media can only be significant in developed countries, where internet penetration is high, or in countries in a democratic setting, where access to the internet is not stifled. However, the power of the convergence of various media, traditional and new, makes this contention somewhat moot. The recent political upheavals in Middle Eastern and African countries — not known for democratic governments and high internet penetration — show how social media can help facilitate information access and communication, and provide a means to mobilise citizens for political activities (Gustin 2011; Sutter 2011). Whether blogs and social networking sites were a catalyst for these series of mass uprisings is subject to much debate (Beaumont 2011; P. N. Howard 2011; Gladwell and Shirky 2011). But such is the power of social media, perceived or real, that autocratic regimes affected by these revolts have resorted to pulling the "internet kill switch" (Cohen 2011) at some point, whilst governments of some democratic countries like the U.S. have pending legislation that mandates government to shut down the internet in cases of cyber security threats (*The Economist* 2011).

The impact of social media, particularly blogs, on politics and civic participation in developed countries has been studied extensively. However, in developing countries like the Philippines, it is surprising that there has been little research on blogging, and almost none on political blogs, even with its liberal democratic setting and highly literate population. Insights on the phenomenon are limited to anecdotes and feature stories that add 'flavour' to the usual news format. It is common fare for TV news programmes to scan blogs, *Facebook,* and *Twitter* for 'public opinion' on a topic, instead of or in addition to interviewing experts, or in the absence of a formal survey. News anchors of a major TV network also spend at least a fourth of airtime reading comments posted on their social networking sites.

Despite low internet penetration, barely 30% of the total population, Filipinos' internet use is growing and diversifying. In 2008, Filipino blog readers reached 3.3 million or 90.3 percent of the country's "active internet universe" — individuals aged 16–54 years going online regularly — a high penetration second only to South Korea (Universal McCann 2008). In 2009, AC Nielsen estimated that one in three Filipino internet users had a blog (Burgonio 2009). Filipino *netizens* are also known for being social networking-savvy. In 2008, the Philippines had the highest social networking penetration, 83.1 percent of its internet users, compared to the global average of 57.5 percent (Universal McCann 2008). Today, the Philippines ranks among the top countries in terms of *Facebook* users, according to the latest figures of Socialbakers (2011) and comScore (Newsbytes.ph 2011).

It is evident that Filipinos actively use, and have great interest in, social media. Whilst there is good logic in government using these tools to engage its citizens, the rationale behind the strategy should be based on a critical understanding of who *netizens* are, their usage of blogs and social networking sites, and whether and how they use social media to engage in politics.

A SIRCA-funded research project, focused on the actors behind the Filipino political blogosphere, was conducted from February 2009 to August 2010. The exploratory study examined A-list Filipino political bloggers and their readers — who they are, why they write and/or read political blogs, their attitude toward politics, and whether and how they access blogs for political participation, both online and offline. A survey was conducted among 30 A-list Filipino political bloggers based on the survey designs of McKenna and Pole (2004), which examined the profile of A-list political bloggers and their online/offline political activities (based on Verba, Schlozman and Brady 1995), and Ekdale et al. (2007), which looked at bloggers' motivation for starting and continuing to write about politics. A survey was also conducted among 64 of the A-list blogs' readers to determine their modes of political participation and political attitude (based on the design of Gorospe-Jamon 1998). The results of these two surveys were triangulated with data from semi-structured elite interviews with top political bloggers, academics, journalists, and communications experts, and focus group discussions (FGDs) involving bloggers and readers in Metro Manila, Cebu (Visayas region), and Davao (Mindanao region).

What follows is a discussion of the study's results and their implications for government-citizen online engagement:

The A-list Filipino political bloggers and their readers are found to be mostly young males, 25 to 34 years old, Metro-Manila based, college-educated, employed, high-income (household earns P50,001 or above, monthly), and veteran internet users (usage of more than seven years and regular broadband connection at home or work). Interestingly, these attributes closely resemble the blogger profile in *Technocrati's State of the Blogosphere 2010 report* (Technocrati 2011). These results imply that active *netizens* who blog about politics and read them are those who have the resources (time, high level of education and income, internet connection, etc.) and are most exposed to national issues due to their geographicl proximity to the seat of political power. This seems to indicate a certain level of elitism and polarisation in Filipino political blogging, where bloggers attract a certain type of audience with a particular background, context, or mindset.

Most bloggers feel that they are already somewhat represented by a political party or interest group, which may explain why most of them have no affiliation to a political group. Top political bloggers are mostly engaged in expressive participation, both online and offline. They use their blogs primarily to "announce an event"; no blogger reported posting "paid advertisement for politicians." Although most bloggers self-rate their site as "critical of government," a majority of them encourages their readers to engage in conventional, lawful forms of political activities. This suggests the dominance of highly politicised *netizens* who lead private lives outside of politics and the presence of critical yet law-abiding, non-hostile actors (moderate left) in the Filipino political blogosphere. In short, these are "Jose Rizal-type" revolutionaries who believe that the pen is mightier than the sword (Jose Rizal is a national hero, an intellectual who wrote poems, essays, and novels to express his aspiration for a free country. His works inspired an armed revolution that led to Philippine independence from the Spanish colonisers in 1898, but which Rizal himself did not believe in). What this also tells government officials and politicians is that political bloggers shun the idea of writing for pay, and any attempt to buy them off might simply backfire.

Credibility is currency in the blogosphere. A-list political bloggers engage and promote each other but show antagonism toward those they identify as writers for politicians or paid hacks. Similarly, blog readers look to credible bloggers, especially those who have established a good reputation in the offline world. Reputation built through exposure in traditional mainstream media is also given premium, hence, the popularity of journalist-bloggers. This suggests that active *netizens* in the political blogosphere have developed a "PR detector." A bad reputation is bad whether online or offline. Blogs and social media peppered with positive press releases and good reviews do not necessarily change perception. If the truth is ugly, adding another layer of public relations masking in cyberspace does not magically morph it into something pretty, especially if the readers are more discerning and inquisitive.

One needs to be cautious, however, in assuming that Filipinos use the internet for political purposes. The same 2009 cyber-campaigning study (Mirandilla 2009) revealed that *netizens* use the internet primarily to access news and current affairs, and information related to work, family, and friends. Information about politics and government was not a top priority even at a time near the election campaign. This suggests that citizens will not actively seek out government officials and politicians in blogs and social media, unless they offer content and services that are relevant or somehow connected to what *netizens* need and want when they use the internet.

The SIRCA study also found that top bloggers started to write about politics mainly to "keep track of their thoughts," a form of self-reflection and expression. Bloggers also write to inform people of the most recent and relevant information about politics. However, they do not necessarily set out to influence their readers or mainstream media through their writing.

Over time, bloggers are motivated to continue writing because it helps them formulate new ideas. These results suggest that actors in the blogosphere are not driven by an advocacy, political ideology, or a grand plan to change the world. They are independent-minded individuals who are motivated to express their political views and, in the process, create new ideas. Sharing up-to-date and relevant information is a critical thread that holds the relationship of bloggers and their readers. This was seen in a few crisis situations, such as when Typhoon *Ondoy* (Tropical Storm *Ketsana*) hit Metro Manila in September 2009. The "unorganised *netizens*," with the support of mainstream media, instantaneously turned blogs and social networking sites, such as *Facebook* and *Twitter*, and *YouTube* into tools for documenting and locating worst-hit areas, survivors, and missing people, and for coordinating relief collection and distribution. It would be prudent for concerned government agencies to learn from that experience by preparing a similar coordinated effort using social media for when the next crisis happens.

Political blogs are undoubtedly a useful medium for acquiring and sharing information, which serve to raise awareness and enrich knowledge about political issues. This increased knowledge and enhanced understanding of "real politics," however, may heighten cynicism and downplay the value of resources that enable participation, as people feel more uncertain about their influence on politics and the impact of their individual actions in the larger scheme of things. This may explain the moderate level of political efficacy among the blog readers surveyed, despite the resources they possess that allow them to participate in politics. In addition, the feeling of efficacy may be mitigated by other critical pre-conditions, such as the absence or presence of opportunities and infrastructure offline, and the effectiveness of institutions to enable citizen participation that will lead to legitimate and effective results. The results also emphasise the need to contextualise the impact of social media and the internet, in general. One must examine how citizenship is traditionally expressed and strengthened in a country with limited resources, whether and how these translate into the online environment, and if innovation is actually supported by real-world institutions.

Overall, the SIRCA study suggests that political blogs have yet to create a tangible macro-impact on how Filipinos participate in politics, considering that the contribution of blogs is concentrated in information-sharing to a small, niche audience, at least for now. Although blogging is generally perceived by its users as having led to an exchange of ideas outside cyberspace, it has yet to transcend appreciation of information and views to more meaningful deliberation. There seems to be no indication that blogs influence political activities outside the blogosphere, at least when it is "politics as usual." As a standalone, its reach is limited to a small community of often like-minded individuals who may already be politicised and engaged to begin with.

Despite its small audience, however, political blogs gain traction and expand its reach when they: (1) go viral on *Facebook* or *Twitter*; (2) get picked up by mainstream media; and/or (3) attract the attention of the gatekeepers. Blogs can spread like wildfire when they get "shared" in social networking sites and are exposed by mainstream media. Blog posts and other messages via social media that make their way to the gatekeepers — individuals who filter information for the decision-makers — also have the potential to contribute to, or influence political discourse, or even the policy agenda.

Otherwise, political blogs serve as a means for citizens to: get up-to-date and more detailed information about a particular issue, access in-depth analysis often lacking in mainstream media, and share their views with others on relevant issues, thus, the ability of blogs to disrupt traditional debate on public issues. Whilst it is difficult to establish whether and how blogs influence political participation, it is clear from the study that this effect is constrained by a blog's reach, the level of cynicism and efficacy of *netizens* accessing blogs, and the vessel that carries the message from cyberspace to the real world.

The current administration of President Benigno S. Aquino III has embarked on what can be called a new governance approach using social media, which is a continuation of his Obama-like campaign strategy in May 2010. President Aquino started official *Facebook* and *Twitter* accounts that are connected to the Official Gazette,[1] with the aim of opening government to feedback from citizens. Whilst it is common for citizens to write letters to the president, it was a historic first when President Aquino himself directly replied[2] to a citizen's note posted on *Facebook*. Indeed, using social media makes the president appear accessible. Although some criticise the move as superficial, the fact that the president uses the same *Facebook* as you and I do has undeniably affected the way in which citizens view their government — a precedent that will change the course of government-citizen communications. Hence, this newly-established government-citizen relationship in cyberspace provides a huge challenge and opportunity for democratic engagement. What is crucial now is how both government and citizens will act at this crucial juncture and emerge from the learning process.

Notes

1. "Official Gazette," last modified 17 November 2011. <http://www.gov.ph/>.
2. "Letter of His Excellency Benigno S. Aquino III, President of the Philippines, in

response to the Facebook note: Mr President, Something In You Has To Die,"
Official Gazette, 13 September 2010. <http://tinyurl.com/33t8qnz>.

References

Beaumont, Peter. "Can social networking overthrow a government?" *The Sydney Morning Herald*, 25 February 2011. <http://tinyurl.com/6kuksa8>.

Bradley, Anthony J. "A new definition of social media." In *Gartner blog network*, 7 January 2010. <http://blogs.gartner.com/anthony_bradley/2010/01/07/a-new-definition-of-social-media>.

Burgonio, Maricel E. "Philippine Internet use rises." *Manila Times*, 27 May 2009. <http://archives.manilatimes.net/national/2009/march/27/yehey/business/20090327bus8.html>.

Campbell, Amy. "Social media — a definition." In *Amy Cambell's web log at Harvard Law*, 21 January 2010. <http://blogs.law.harvard.edu/amy/2010/01/21/social-media-%E2%80%94-a-definition>.

Cohen, Noam. "Egyptians were unplugged, and uncowed." *The New York Times*. 20 February 2011. <http://www.nytimes.com/2011/02/21/business/media/21link.html?_r=1&emc=tnt&tntemail1=y>.

Cooper, Abraham and Harold Brackman. "Why the internet is a double-edged sword." *The Globe and Mail*, 2 July 2009. <http://tinyurl.com/6zz4543>.

Economist, The. "Internet blackouts: Reaching for the kill switch — the costs and practicalities of switching off the internet in Egypt in elsewhere." *The Economist*, 10 February 2011. <http://www.economist.com/node/18112043>.

Ekdale, Brian, Kang Namkoong, Timothy Fung, Muzammil Hussain, Madhu Arora and David Perlmutter. "From expression to influence: Understanding the change in blogger motivations over the blogspan." *Association for Education in Journalism and Mass Communication Conference*, 2007, 8 May 2009. <http://www.journalism.wisc.edu/smad/papers/ekdale-namkoong-fung-hussain-arora-perlmutter-2007.pdf>.

Gladwell, Malcolm and Clay Shirky. "From innovation to revolution: Do social media make protests possible?" In *Foreign Affairs*, March/April 2011. <http://tinyurl.com/65rxa8o>.

Gorospe-Jamon, Grace. "The El Shaddai prayer movement: A study of political socialisation in a religious context". Ph.D. dissertation, University of the Philippines, 1998.

Grossman, Lawrence. *The Electronic Republic*. New York: Penguin Books, 1995.

Gustin, Sam. "Social media sparked, accelerated Egypt's revolutionary fire." In *Wired*, 11 February 2011. <http://www.wired.com/epicenter/2011/02/egypts-revolutionary-fire/>.

Hacker, Kenneth L. "The White House computer-mediated communication (CMC) systems and political interactivity." In *Digital democracy: Issues of theory and practice*, edited Kenneth L. Hacker, and Jan A.G.M. van Dijk. Thousand Oaks, SA: Sage, 2000.

Howard, Alex. "Are the internet and social media 'tools of freedom' or 'tools of Oppression'?" In *gov20.govfresh*, 8 March 2011. <http://t.co/aHvroS6>.

Howard, Philip N. "The cascading effects of the Arab Spring." *Miller-McCune*, 23 February 2011 <http://tinyurl.com/3aw4gtc>.

Iowa State University. "What does 'new media' mean?" <http://newmedia.engl.iastate.edu/about/what_is_new_media>.

Kelly, Sanja and Sarah Cook, eds. *Freedom on the net 2011: A global assessment of internet and digital* media, 18 April 2011. <http://www.freedomhouse.org/template.cfm?page=664>.

Kenski, Kate and Natalie Jomini Stroud. "Connections between internet use and political efficacy, knowledge, and participation." *Journal of Broadcasting & Electronic Media* 50 (2006): 173–92.

Manovich, Lev. *The language of new media*, 2001. Cambridge, Massachusetts/London, England: MIT Press. <http://tinyurl.com/yb6db2u>.

McKenna, Laura and Antoinette Pole. "Do blogs matter? Weblogs in American politics." Paper presented at the annual meeting of the American Political Science Association, Hilton Chicago and the Palmer House Hilton, Chicago, IL, 2 September 2004. <http://www.allacademic.com/meta/p60899_index.html> (accessed 5 June 2009).

Mirandilla, Mary Grace P. "Cybercampaigning for 2010: The use and effectiveness of websites and social networking sites as online campaign tools for the 2010 presidential election." A research funded by a grant from the Philippine ICT Research Network with the aid of IDRC Canada, August 2009. <http://papers.ssrn.com/sol3/papers.cfm?abstract_id=1553724>.

———. "A-List Filipino Political Bloggers and Their Blog Readers: Who They Are, Why They Write/Read Blogs, and How They Perceive and Participate in Politics." *Media Asia Journal* 38 (2011): 3–13.

Newsbytes.ph. "PH ranked as top social networking nation." *Newsbytes.ph*, 25 March 2011. <http://tinyurl.com/66vh7vo>.

Nisbet, Matthew C. and Dietram A. Scheufele. "Internet use and participation: Political talk as a catalyst for online citizenship." *Journalism and Mass Communication Quarterly* 81 (2004): 877–96.

OpenNet Initiative. *2010 Year in Review*. <http://opennet.net/about-filtering/2010yearinreview>.

Putnam, Robert D. *Bowling alone: The collapse and revival of American community.* New York: Simon and Schuster, 2000.

Singer, Jane B. "The political j-blogger: 'Normalizing' a new media form to fit old norms and practices." *Journalism* 6 (2005): 173–98.

SocialBakers. *Philippines Facebook statistics*, 2011. <http://www.socialbakers.com/facebook-statistics/philippines>.

Sutter, John D. "The faces of Egypt's 'Revolution 2.0'." *CNN.com*, 11 February 2011. <http://tinyurl.com/66vc8v6>.

Sweetser, Kaye D. and Linda Lee Kaid. "Stealth soapboxes: political information efficacy, cynicism and uses of celebrity weblogs amongst readers." *New Media & Society* 10 (2008): 67–91.

Technocrati. *WHO: bloggers, brands and consumers — day in state of the blogosphere 2010 report*, 17 March 2011. <http://technorati.com/blogging/article/who-bloggers-brands-and-consumers-day>.

Tedesco, John C. "Examining Internet Interactivity Effects on Young Adult Political Information Efficacy." *American Behavioural Scientist* 50 (2007): 1183–94.

Universal McCann. *Wave.3 Power to the People — Social Media Tracker*, 12 October 2008. <http://universalmccann.bitecp.com/wave4/Wave4.pdf>.

Verba, Sidney, Kay Lehman Schlozman and Henry Brady. *Voice and equality: Civic voluntarism in American politics*. Cambridge, Mass.: Harvard University Press, 1995.

12

INTEGRATING DIGITAL AND HUMAN DATA SOURCES FOR ENVIRONMENTAL PLANNING AND CLIMATE CHANGE ADAPTATION
From Research to Practice in Central Vietnam

Pham Huu Ty, Richard Heeks and Huynh Van Chuong

ABSTRACT

Climate-related and climate change-related events are causing problems for agricultural communities in many developing countries, and are requiring new practices that enable these communities to adapt. In planning adaptation practices, two quite different data sources have been common — top-down digital sources such as those provided by remote sensing, and bottom-up human sources such as those provided by participatory events. These can be brought together via geographic information systems to produce a more robust guide for adaptational decision-making. This paper reports one such initiative in Central Vietnam which integrated local and external data sources

— geographical, economic, political, and agricultural — in order to provide guidance on environmental planning. The paper reports on the transfer from research modelling to environmental practice, and also draws some broader conclusions about data source integration for ICT4D practitioners including the need to re-engineer planning processes rather than simply automate traditional — often narrowly-scientific and top-down — approaches.

ICTs AND CLIMATE CHANGE ADAPTATION STRATEGIES

Although there is still ongoing debate, there seems to be a general consensus that climate change is a reality, and a reality that is causing particular problems for farmers in developing countries (Ospina and Heeks 2010a). There are problems such as the melting of glaciers, flooding, and spread of desertification or of new vector-borne diseases due to changes in temperature (IPCC 2007). In this paper, we will focus on two problems specific to highland farming: landslides and soil erosion.

Of course, these have occurred throughout the history of agriculture, with landslides destroying life, livestock and property as well as agricultural land, and with erosion of two types — soil run-off from gullies and loss of land from river-sides — both impacting availability of good quality agricultural land (Chung et al. 1995; Morgan 2005). However, there is mounting evidence that both landslides and erosion are increasing and will further increase as a result of climate change, particularly due to changes in precipitation patterns (IPCC 2001; Ray and Joshi 2008).

Adaptation to climate change is therefore required, and this will include environmental planning actions that address the increased likelihood of landslide and erosion events that create problems for highland agriculture. The whole notion of adaptation to climate change is only just starting to emerge onto the policy agenda in developing countries (Ospina and Heeks 2010b). As it emerges, it has been guided by prior approaches to environmental planning. But such approaches have been significantly criticised on two key shortcomings (Weiner et al. 2002). First, they lack local ownership and participation. Second, they rely on too narrow a range of data sources, often external, scientific data sources relating to physical geography.

There has been growing use of Information and Communication Technologies (ICTs) in both environmental planning and climate change adaptation strategies; indeed, ICTs have become an essential component of such strategies (Adele and Grazia 2005; Ospina and Heeks 2010b). However, the potential for ICTs to address the problems of local participation and data source integration has yet to be realised. In most cases, ICT-based strategies merely replicate the top-down, narrowly-scientific approaches of earlier non-ICT-based equivalents (Adele and Grazia 2005; Sarkar et al. 2008).

The aim of this research, therefore, was to try a different approach; investigating how ICTs could be used for local environmental planning in the context of climate change, and used in a way that enabled local participation and the integration of a relatively broad range of data sources. As described next, the chosen location was in central Vietnam.

CLIMATE CHANGE ADAPTATION AND LAND DEGRADATION IN VIETNAM

Vietnam was the chosen study site. Climate change is already recordable in the country, using both scientific historical records and the testimony of local people. These show an annual average temperature increase of 0.7°C from 1951 to 2000, and a rise in average sea level of roughly 20 cm (MONRE 2008). Overall changes in precipitation levels have been less clear, largely because they show greater local variations. However, there is evidence of greater year-to-year fluctuations in rainfall levels, a more prolonged rainy season, and — of particular relevance — evidence of greater numbers and greater intensity of rainstorms (Dien and Trang 2008). This, in turn, has led to evidence of increased weather-related damage in rural areas including landslides and erosion. These are predicted to increase in future as climate change increases (MONRE 2009).

Such damage is particularly problematic for Vietnam. Despite only 19% of GDP now being produced by agriculture, it is responsible for more than 55% of employment and rural areas are home to over 70% of the population (*The Economist* 2008). Vietnam is also particularly susceptible to rain-induced damage to land because three-quarters of its land area is mountainous with high rainfall, steep slopes and readily-erodible topsoil. It is thus estimated that some 17.7 million hectares of land (roughly half the total land area of the country) is at risk of rain-induced degradation (VEA 2005). These pressures have been exacerbated by the *doi moi* liberalisation policies in place since 1986 which have had the effect of placing economic above environmental considerations in guiding both national and local agricultural policies and practices (Dien and Trang 2008; Scott and Conacher 2008).

That is not to say that there has been an absence of strategies, with the Ministry of Natural Resources and Environment at the national level recognising the importance of environmental planning and, more recently, of climate change adaptation, and building these into its statements and plans (e.g. MONRE 2003; MONRE 2008). However, there have been serious shortcomings in the operationalisation of environmental and adaptational planning (as is the case in many developing countries) (Phuong 2008). Participation of local people has been negligible. As a result, the ownership of such plans and the inputs of local knowledge and data have often been poor. This in turn leads to plans that are unworkable in practice. The top-down approach that is common typically involves the use of some national, or at best, provincial data, interpreted through the lens of local political and economic interests as determined by government officials. Lack of objectivity and rationality in the whole process is apparent. There has been some rhetoric of participation but this has not been matched by reality (Scott and Conacher 2008). There are examples of ICTs — especially Geographic Information Systems (GIS) — being introduced for environmental purposes but in ways that are neither participative nor integrative of wider data sources (Van De et al. 2008).

Vietnam was therefore seen to provide a microcosm of patterns in developing countries — of land degradation exacerbated by climate change; of planning processes that were too top-down and based on too narrow a foundation of data; and of a largely as-yet-unrealised potential for ICTs to make a contribution. Further details now follow of an attempted use of ICTs to contribute to environmental planning in a context of climate change, via a process that was participative and data-integrative.

THE CASE STUDY LOCATION

Thua Thien Hue province in Central Vietnam was chosen as the test site for use of a participative, integrative environmental planning approach based around a GIS and in the context of climate change. This province is typical of areas facing increasing climate change problems with landslides and erosion. It is characterised by steeply-sloped mountains and hills (upland areas account for more than 70% of its total area) that have been particularly susceptible to erosion due to the patterns of land use (Kiet 1999). The local population — with poverty rates nearly twice the national average, and largely consisting of ethnic minorities — are thus more vulnerable to climate change than most; and their agricultural vulnerabilities have been increased with the construction of hydroelectric plants that cause deforestation, in turn increasing land degradation (De Koninck 1999; C-Core 2010).

Within the province, our focus was the district of A Luoi, which houses the Ta Rinh watershed, an area reporting significant problems with landslides and soil erosion. The area has a population of nearly 21,000 in which more than 80% are dependent on incomes from either agriculture or forestry (more than half the land is wooded), but using agro-forestry systems that are largely traditional and of low productivity. That dependency reduces their resilience and increases their vulnerability to climate change.

INTEGRATED ICT-ENABLED PLANNING IN PRACTICE

Six steps were applied in this research in order to enable better environmental planning and an adaptive response to the threat of climate change.

Step 1: Participatory GIS Mapping of Soil Erosion and Landslides

We began with a combination of top-down and bottom-up data gathering. First, all available secondary data were collected from different departments in A Luoi district, including land use, rainfall, and other socioeconomic data plus a base map, all of which were loaded onto the GIS (Arcview 3.3). Then the first meeting was held with representatives of the district Rural Development and Natural Resources and Environment departments, and of the nine communes (communes are the smallest governance areas; A Luoi district has a total of 13 communes and the nine in the Ta Rinh watershed area have been included in the study; those representing communes had a dual role as both commune representative and farmer). They were asked particularly about soil erosion and landslide status in their area.

It was reported that upstream rivers in the Ta Rinh watershed area have been facing severe erosion along their banks. Each commune has been losing at least three hectares of productive land area annually; several communes have lost up to five hectares. Observations showed that most of the land lost either had crops (such as rice) under cultivation or was productive forest (rather than unused land). Erosion was thus negatively impacting farmer income and food security. Riverside erosion was also threatening some residential areas, leading to a need for some farming households to be resettled. Figure 12.1 provides a summary of erosion events as recorded on the GIS.

Fifteen landslide sites were identified by local people, with most occurring on steeply-sloped land with little cover or with plantation forest of young trees. The most significant landslide had removed half a hectare of forest, damaging local houses and also posing a threat to life. We also asked about erosion within hill-side gullies. These occurred mainly where land had little cover or just young forest, but such erosion was not rated by the local farmers to be as important as river-side erosion and landslides.

FIGURE 12.1
Erosion and Landslide Map

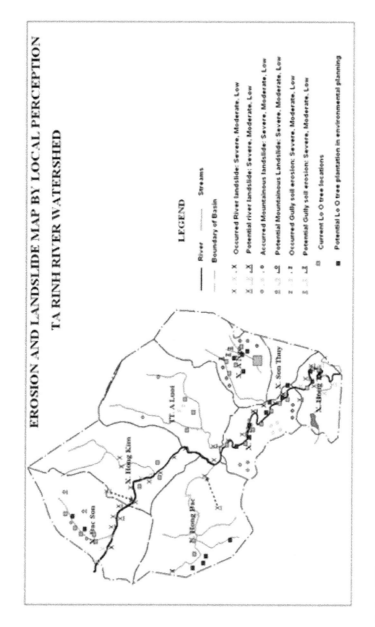

Following this preliminary mapping, reported locations of erosion and landslide were visited and checked via GPS tool. The data were then combined with prior knowledge of risk factors from the scientific literature and from the local participants to produce a mapping of likely risk areas for future erosion and landslides (i.e. those areas deemed most vulnerable to climate change-related increases in rainfall intensity). With the local participants, we used an 'analytical hierarchy process' approach, which got them to paired compare each risk factor in turn to all others, and rank them by priority weight.

Then, finally, we also asked directly about climate change, although this was not mapped onto the GIS. Alongside reports of changes in temperature and seasonality (e.g. with longer and more severe cold periods that damaged crops), it was reported that there were also changes in precipitation patterns with greater rainfall during the rainy season and a greater intensity of rainfall that impacted both the number and scale of river-side erosions and landslides.

Step 2: Viability Testing of Community-based Adaptation Options

One of the key shortcomings of current planning has been the potential unsuitability of proposed adaptational options. A central part of the new approach was therefore to test out the viability of such options prior to proposals being made. As an example, a typical adaptational response around river erosion has been the construction of artificial walls, dykes and channels. However, the developmental objectives and mindset of the district government made this inappropriate: they were unwilling to spend money on such construction and saw higher priorities in spending on more direct contributions to socioeconomic goals such as agricultural improvement, education, healthcare and transportation infrastructure. Discussion groups organised with farmers in each of the nine communes showed a similar resistance to money being spent on construction work. They, too, prioritised spending on other development objectives.

More aligned with the political economy of the area was a proposal to plant a particular type of bamboo (known locally as Lo O: see Figure 12.2). The district government already had a plan to delegate responsibility to each commune for development of a bamboo plantation, but had not implemented it due to lack of guidance on where would be best to locate the plantations.

In the discussion groups, farmers were also favourably disposed towards this adaptational option since this would be a farming-based technique, since

FIGURE 12.2
Lo O Bamboo Growth, Harvesting and Use

River bank protection   Trunks waiting for selling   Bamboo chopstick, toothpicks manufactures

Effective protection for dry crops   Lo O tree delivery in Hue City   Goods from Lo O tree trunks

Source: Pham Huu Ty.

some already used small Lo O plantations to protect their crops or houses, and since they were already aware that there was some income potential from selling bamboo. Despite this support from both government and farmers, however, we felt it important to gather economic data in order to assess the financial viability of this option, given prior occasions on which adaptational options had been proposed that proved to be not economically viable (Phuong 2008).

The economic data was gathered via a field survey which interviewed farmers, traders, and bamboo processing factory owners. The survey showed that each commune already had an average area of 2–4 hectares of Lo O bamboo under cultivation, but that the number of trees had fallen dramatically: from over 45,000 in 2006 to less than 20,000 in 2008. However, this proved to be due to positive rather than negative aspects of the tree. Demand for the bamboo had been — and was still — high due to its use in producing toothpicks, chopsticks, furniture, farming tools and other items. As a result, prices were high and rising; thus inducing those with access to bamboo to sell. As can be seen from Table 12.1, both the sales price for farmers locally but also the price achieved by traders in local cities had gone up considerably. Consumer price inflation — 45% from 2006 to 2009 — accounted for some of this increase, but with sales prices rising by around 100% during the same period, profitability from Lo O bamboo had definitely risen during recent years, with farmers estimating they could earn at least 2 million dong (c.US$100) per hectare (though many had earlier held back from doing so due to the 3–6 year period required for growth before harvesting would be possible).

TABLE 12.1
Bamboo Sales Prices for Farmers and Traders, 2006–09

Year	2006	2007	2008	2009
Sale price for farmers	5,000	6,000	8,000	1,0000
Sale price for traders	10,000	15,000	17,000	21,000

Unit: VND per tree (US$1=c. VND 21,000 in 2007/08 and at the time of writing in 2012)
Source: Field surveys

Lo O bamboo is also known to have other benefits. It is very easy to plant, and can be grown by putting a piece of trunk into the soil to develop another tree without any fertiliser. Most farmers proved to have a good awareness of how, where and when to plant the bamboo. For adaptational purposes of environmental protection, it is good because of its large and deep root system, and those farmers who had used Lo O reported the bamboo variety to hold soil well and to protect land and property from water-borne erosion and landslide.

In all, this type of bamboo was shown to be politically, economically, agriculturally, and adaptationally viable.

Step 3: Adaptational Planning

Hong Kim commune in the Ta Rinh watershed was selected for the next stage of the process, which would be a specific action plan for Lo O bamboo planting. Many ICT-based methods at this stage use the technology to present an "answer" and then limit participation to its presentation to local citizens. We wanted to follow a more participative and bottom-up process. We, therefore, held several meetings with local groups in the commune to get them to guide us on the most appropriate locations for planting, using their indigenous knowledge as expressed in a set of criteria.

As summarised in Table 12.2, farmers said that in order to control erosion and landslides, they typically planted Lo O trees within 5–10 metres of the edges of rivers and streams, and on sloping land within 10–20 metres of the back of their houses. Alluvial deposit areas along the rivers emerged as a priority for planting the bamboo because they are generated by soil deposit and, having low soil quality, are not suitable for crops. In addition, Lo O trees were identified as locatable within five metres along hill bases to manage mountainous erosion and landslides.

TABLE 12.2
Summary of Criteria for Lo O Bamboo Plantation Planning

Location	Water edge	Hillside (property)	Alluvial deposit	Hill base
Distance	5 – 10m	10 – 20m	On deposit	5m

Step 4: GIS Model Building

We did not want to make an assumption that all four planting locations should be equally weighted. We therefore used the 'analytical hierarchy process' approach once again to allow the farmers to rank them by preference. In practice, the four locations emerged as equally ranked. As summarised in Figure 12.3, these four were input with equal weighting into the GIS, utilising four map layers — river system, housing, land use, and topology — which were collected together with the base map to operate a binary model ('true' for those areas that met the logical expression based on the location criteria, 'false' for those areas that did not) for selection of locations for Lo O bamboo planting. This was then combined with the original map data on erosion and landslide risk areas, to produce a set of prioritised planting areas.

Step 5: Physically Checking the Suitability of the Lo O Bamboo Plantation Map

After producing the Lo O plantation map (see Figure 12.4), it was re-checked with local farmer groups in Hong Kim commune to determine its

FIGURE 12.3
Conceptual Framework of GIS Model for Bamboo Planting

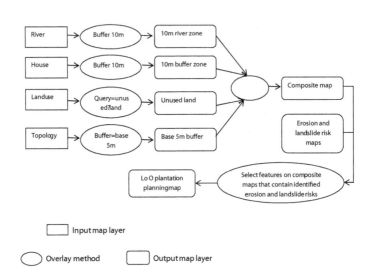

Source: Authors

appropriateness. First, the discussion was based on observations on produced maps, then the research group and local farmers went to several sites which were identified for Lo O planting. Most planning sites that had been selected along the river banks and alluvial deposit areas proved to be suitable because they were well-defined on the maps (and, indeed, some areas — as previously known and indicated in the original map — already had some bamboo planted). However, other Lo O plantation areas — mainly those along the hill bases and behind local houses — were found in practice not to be appropriate: due to limitations in the quality of topographical and housing data. One conclusion, therefore, was the need for better data, and also the need to better analyse the topology map in order to define hill or mountain bases more accurately.

Step 6: Making the Final Adaptation Plan

The final physical check was incorporated into the results of the GIS model building, and a final adaptation plan was completed by the research group and shared at a closing workshop of local government and farmer representatives.

FIGURE 12.4
Lo O Plantation Planning Map in Hong Kim Commune

Source: Project GIS.

In general, the final plan was well-received — there was a strong awareness that the ICT component had made an invaluable contribution, but also that the input of local data and the participation of local people was a key necessity in achieving both the accuracy and the ownership of the final plan which they would proceed to implement. The management of the process became a central point of discussion, with the key proposal that — rather than leaving these adaptational actions to individual farmers (who, it seemed, had so far not been guided by market prices to invest heavily in *Lo O* plantations) — it would be better to allocate plantation as a community responsibility, much as already happens with community forest allocation, something with a long history and clear legal framework within Vietnam.

Ideally, local ownership for future adaptational planning activities could have been enhanced by handing over the ICT-based components to some representatives within the local community. However, those present felt that this should remain the responsibility of outside experts, who cannot be completely done away with owing to the skill, knowledge and technology requirements associated with GIS usage.

IMPLICATIONS FOR ICT4D PRACTICE

Climate change is on the increase, and the planning of adaptational actions increasingly finds a place in the agenda of developing countries. ICTs can potentially have a key role to play in climate change adaptational strategy. However, our experience in traditional environmental planning practices in this case suggests that simply using ICTs to "pave the cow-paths" — that is, to automate existing top-down, scientific processes of decision making — may achieve little.

Instead, alongside the introduction of digital technologies, there needs to be a re-engineering of environmental planning processes. There has long been recognition of the value of a more bottom-up, participatory approach to such processes. However, as noted earlier, this has too rarely been translated into reality or, if translated, has been paid only lip service, with output from ICT-based decision-making merely presented for limited discussion to local representatives.

We wanted — indeed, felt it essential — to go well beyond this, to a truer process of local participation and greater use of a wider range of local data sources. Thus we ensured local representatives were used to provide baseline data on vulnerabilities, data on both the politics and economics of potential adaptational actions, criteria for the specifics of those actions, and a cross-checking of suggested actions provided via the ICT system. Throughout, there was also intensive involvement in discussions of the nature of climate change, its impacts, potential solutions, and implementation.

This is a time-consuming and costly approach. And one in which ICTs ultimately play a key role but only one role amongst many, alongside other technologies, other actors, and other knowledge. However, we believe such an approach is central if local communities are to effectively adapt to the agricultural and other climate-related challenges that they are likely to increasingly face in the future.

References

Adele, Celino and Concilio Grazia. "Open content systems for e-governance: the case of environmental planning." Paper presented at eGOV05 Workshop, 13 September 2005. United Kingdom: Brunel University.

C-Core. *Water management service trial.* Kanata, ON: C-Core, 2010.

Chung, Chang Jo F., Andrea G. Fabbri, AG., and Cees van Westen. "Multivariate regression analysis for landslide hazard zonation." In *Geographical information systems in assessing natural hazards,* edited by Alberto Carrara and Fausto Guzzetti, 107–33. Amsterdam: Kluwer Academic, 1995.

De Koninck, Rodolphe. *Deforestation in Viet Nam.* Ottawa, ON: IDRC, 1999.

Dien, Le Thien and Dang Ngoc Kim Trang. "Climate change adaptation of local people in Nam Dong District, Thua Thien Hue Province." In *Understanding policy and practice: studies of livelihoods in transition,* 63–84. Hue, Vietnam: Hue University Publishing House, 2008.

Economist, The. "From basket case to rice basket." *The Economist,* 24 April 2008.

IPCC. *Climate change 2001: Impacts, adaptation and vulnerability; Report of the Intergovernmental Panel on Climate Change.* Cambridge, UK: Cambridge University Press.

———. *Climate change 2007: Impacts, adaptations and vulnerability; Introductory Contribution of Working Group II to the Fourth Assessment Report of the Intergovernmental Panel on Climate Change.* Cambridge, UK: Cambridge University Press.

Kiet, Ho. "Soil erosion and deposition assessment on cropping systems in Huong River basin, Thua Thien Hue province, Vietnam." PhD diss., Hanoi University of Agriculture, 1999.

Morgan, Royston Philip Charles. *Soil erosion and conservation.* Oxford, UK: Blackwell, 2005.

MONRE. *Viet Nam initial national communication — under the United Nations Framework Convention on Climate Change.* Hanoi: Ministry of Natural Resources and Environment, 2003.

———. *National target to response to climate change.* Hanoi: Ministry of Natural Resources and Environment, 2008.

———. *Climate change scenario in Vietnam — temperature rise.* Hanoi: Ministry of Natural Resources and Environment, 2009.

Ospina, Angelica Valeria and Richard Heeks. *Unveiling the links between ICTs and climate change in developing countries.* Manchester, UK: Centre for Development Informatics, University of Manchester, 2010. <http://www.niccd. org/ScopingStudy.pdf>.

Ospina, Angelica Valeria and Richard Heeks. *Linking ICTs and climate change adaptation*. Manchester, UK: Centre for Development Informatics, University of Manchester, 2010. <http://www.niccd.org/ConceptualPaper.pdf>.

Phuong, T.T. "Land use planning: policy and practice." In *Understanding policy and practice: studies of livelihoods in transition*, pp. 123–42. Hue, Vietnam: Hue University Publishing House, 2008.

Ray, Prashant Kumar Champati and D.D. Joshi. "The effect of climate change on geomorphic processes and landslide occurrences in Himalaya." Paper presented at SAARC Workshop on Climate Change and Disaster, Kathmandu, 21–22 August 2008.

Sarkar, Shantanu, Debi Prasanna Kanungo, A.K.Patra and Pushpendra Kumar. "GIS-based spatial data analysis for landslide susceptibility mapping." *Journal of Mountain Science* 5 (2008): 52–62.

Scott, Steffanie and Arthur Conacher. "Land degradation and poverty." *Geographical Research* 46 (2008): 1–3.

Van De, Nguyen, Ian Douglas, Julia McMorrow, Sarah Lindley, Dao Kim Nguyen Thuy. Binh, Tran Thi Van, Le Huu Thanh, and Nguyen Tho. "Erosion and nutrient loss on sloping land under intense cultivation in Southern Vietnam." *Geographical Research* 46 (2008): 4–16.

VEA. *State of environment report*. Hanoi: Vietnam Environment Administration, 2005.

Weiner, Daniel, Trevor M. Harris and William J. Craig. Community participation and geographic information systems. In *Community participation and geographic information systems*, edited by William J. Craig, Trevor M. Harris and Daniel Weiner, pp. 3–16. London: Taylor & Francis, 2002.

13

THE CHALLENGE OF WORKING ACROSS CONTEXTS AND DOMAINS
Mobile Health Education in Rural Cambodia

John Traxler

ABSTRACT

In general terms, the academic visibility and professional progress of researchers is a process of acculturation and assimilation into specific communities of established researchers and their discourses. Emergent researchers, especially those outside the mainstream of academic institutions and the wider global research community face two problems. Firstly, they are unaware of many of the tacit or assumed expectations, standards and criteria of the global research community. Secondly, this global research community is not monolithic and undifferentiated but is instead composed of many different smaller, and almost mutually exclusive and almost mutually incomprehensible communities. We explore some aspects of this dilemma from the perspective of a specific SIRCA project faced with the discourses of several research communities, those of software development, mobile learning, health education and development studies, and a mentor coming from his own research communities. It becomes apparent that emergent researchers coming from a position of relative isolation

147

need considerable support in deciding the optimal affiliations and associations for their professional development.

INTRODUCTION

Over the last three years, a team at the Royal University of Phnom Penh, Cambodia, working with support and funding from the IDRC SIRCA programme delivered through the Nanyang Technological University's SiRC in Singapore, has worked on a project to enhance health education — specifically sexual and reproductive health education — for rural Khmer youth, exploiting the affordances and ownership of messaging on their low-specification mobile phones.

The explicit focus of the project in Cambodia and of the programme as a whole has been building research capacity in the poorest countries of South Asia but in fact this specific project and others of its kind stand at the intersection of several different and perhaps divergent or competing discourses. These include the discourses of:

1. Mobiles-for-health (mHealth), part of a wider mobiles-for-development (m4d) discourse
2. ICT for 'development' (ICTD), part of a wider discourse around both software and hardware technology development and also the problematic concept of 'development' itself
3. Mobile learning, a discourse drawing on largely discrete research endeavours based ideologically in the separate education systems of Europe, North America and Asia Pacific
4. The sociology of mobilities, a very new discipline that seeks to document and theorise the impact of mobiles and mobility on social practices and activities.

Each of these discourses comes from its own relatively coherent and relatively hermetic communities with its own heritage, language, literature, paradigms, aspirations and its own philosophical foundations.

In general terms, academic visibility and professional progress require project workers such as those in the team in Phnom Penh to make claims for their work around five central issues, namely that this work is:

1. Unique, in the sense that it has not been done before, in however small a respect, and that this claim is based on a review of the appropriate literature;

2. Relevant and meaningful, perhaps in that it is generalisable or transferable and thus also valuable and significant;

3. Rigorous, in that it shows the application of theory, methods, evidence, data, analysis, inference and discussion to an agreed standard;

4. Repeatable, in providing the guidance, information and resources in sufficient detail to enable other people to replicate or extend the project's outcomes and conclusions or to build on them in some way in a different context; and

5. Ethical, in that *no harm* is done by the research and perhaps things get better.

Whilst these precepts look simple, universal and self-evident, they are in fact highly contextualised and not easily commensurable; they require or assume that research workers are situated in a specific and established research community and have assimilated and understood the basic culture of that community, much of which may be tacit or implicit. We should notice that each of these discourses takes place in its own language and also notice that they were all born towards the end of the 20th century. Their communities, ideas and practices are consequently still fluid and evolving.

We explore the implications of these precepts for the Cambodian project from the perspective of these different discourses. It is however a very brief exploration that only hints at the complexity of the issues.

Cambodian Higher Education Briefly

We must start however by looking at the Royal University of Phnom Penh and the Cambodian higher education environment more generally. Many universities in the developing regions of South Asia have research capacity challenges but we can quickly identify the nature of the specific constraints facing the IDRC-funded team based in the IT Centre. The Khmer Rouge regime and the era of the Democratic Republic of Kampuchea were defining episodes in the recent history of Cambodia. Their impact on the universities of Cambodia and their staff was traumatic. In the words of the University website, "Between 1975 and 1979, the Khmer Rouge regime forced the cessation of formal education. Schools and universities were closed and destroyed, and teaching services decimated. Along with all other educational institutions in Cambodia, Phnom Penh University ceased to function during this time. The Khmer Rouge targeted the educated, and many of the University's staff were killed. Of the educated people who survived the regime, few chose to remain in Cambodia once its borders reopened. Deserted for almost five years, the campus became another victim of the grim civil war which followed the immediate toppling of the Pol Pot regime." This was about 30 years ago and the impact must still be apparent in capacity at the highest levels. The University now has over 420 full-time staff. All of its 294 academic staff hold tertiary qualifications, including 15 PhDs and 132 Masters degrees. PhDs consequently compose only about 5% of the teaching staff. Furthermore, postgraduate courses date from only 2001. The IT Centre hosting the project seems to draw on the local IT professionals for teaching support and many of them have come from these courses. There are obviously enormous challenges building teaching capacity and research capacity in terms of human resource development but also in terms of physical infrastructure. Furthermore, the University organisational chart makes it clear that the IT Centre is distinct and separate from the teaching wing housed under the Department of Computer Science.

It is clear that by the standards of the wider world, the Royal University of Phnom Penh is still very new and is not academically mature. It must negotiate an identity from amongst the models of the so-called *Confucian* universities, the English *Oxbridge* universities, the US universities and various other archetypes.

Ownership of laptops and mobile phones may be widespread amongst students and staff but much more time is needed to establish or re-establish institutional infrastructure, accommodation and estate, and academic culture, procedures and practices. The absence, for example, of staff with doctoral qualifications, of a postgraduate research community, of a formal ethics procedure, the lack of research seminars, the uneven English language skills and the over-reliance on fresh graduates for teaching duties that demand much experience are all indicative of possible weaknesses.

The achievements and challenges of the team in Phnom Penh illustrate the problems faced by new and emergent researchers as they seek an identity and place themselves amongst the various possible research discourses and communities.

THE DISCOURSES OF ICTD

The ICTD community is defined by its attempts to deploy and explore, both in practice and in theory, the use of ICT for *development*. This is not the place to rehearse the difficulties implied by this definition; they include differing definitions of ICT, ranging from conventional computer-based technologies to radio, TV and landline telephony; uncertainty about the nature of *development* and how to promote and measure it, especially once purely material or economic metrics are discarded; the recognition that all (ICT) technologies embody ideologies; the tensions between *appropriate* technology and *development*; the implicit privileging in addressing disadvantage and deprivation of a definition based primarily on geography rather than gender, age, socioeconomic class, and so on.

This is however a sophisticated analysis for emergent researchers to understand and they could be forgiven for uncritically adopting their own preferred technology for attacking what are urgent and pressing social needs. Objective and physical issues such as infrastructure and hardware are often challenging enough without the added complication, for example, of engaging with users and stakeholders and thus complicating the process further whilst demonstrably delaying addressing stark and immediate human needs. A recent review, "*What are the key lessons of ICT4D partnerships for poverty reduction? A Systematic Review Report*" in February 2011 by Marije Geldof, David Grimshaw, Dorothea Kleine and Tim Unwin for DFID, reveals the complexity of *development* interventions in ICT especially for emergent researchers who lack the necessary experience and context.

For researchers in South Asia amongst others, there is the additional complication that the discourses of ICTD talk of *the Global North* and *the Global South* at the expense of any discourse in terms of *East* and *West*, making conceptualising their own experiences still more difficult.

The team in Phnom Penh are working in a *developing* country, one amongst the poorest in the world, and they are working with SMS, the ICTD technology *par excellence*. Although this makes them ICTD practitioners, does it make them ICTD researchers? A key issue here is that of understanding transferability. Just because they are using a universal technology to address a universal problem, does that make their work transferable? Because they are solving a practical problem, will this guarantee their work is rigorous? If the team are to participate in the discourses of the ICTD community, these are the kind of issues that would move them from practitioners to researchers. They are however very new to the skill set of social research that would access stakeholder views and user participation, and not experienced in recognising and critically reading authoritative accounts of similar projects.

The Discourses of the Sociology of Mobilities

This may seem to be a long way from the concerns of health education in rural Cambodia but the team are seeking to understand and exploit mobile phone ownership and use amongst young people, to align their system alongside young people's attitudes, expectations and behaviour with the use of mobiles in their lives; this emergent area of research provides insights and methods into these very issues and provides model accounts of investigating the role and significance of mobiles in communities and sectors around the world. It also explores the demographics and social trends of mobiles, again a possibly valuable resource.

It draws on the older discipline of sociology, within which one might include ethnography and anthropology. Access to this discourse would enable the team, in their interactions with the users of the system, to move beyond the naïve beliefs that, for example, users always say what they mean and mean what they say, and the equally naïve belief that asking users questions *naturally* provides answers that are useful and trustworthy. An increased awareness of social science methods would give the team increased methodological sophistication (and avoid resorting uncritically to questionnaires, focus groups and interviews) and sensitise them to issues such as *interviews-as-performance*, and contrast between *front-of-stage* and *back-stage* accounts and to the issues of sampling, confidence, corroboration and triangulation when working with users and data.

The discourses of social sciences might have provided the most appropriate language for ethical discussions about researcher interventions, and the nature of harm and specifically embarrassment and shame in the project's target community. The overt rationale for the project was indeed perceptions of embarrassment amongst Cambodian young people especially in seeking guidance or information on sexual matters.

The Discourses of mHealth

Using mobiles for health is a well-established technology amongst the *mobile-for-development* community, the *m4d* community, composed mostly of activists, NGOs and practitioners intent on dealing with practical problems and pressing needs. Their discourses are around problem-solving and service delivery. One major focus is South Africa but this raises the issue of relevance and transferability. The Cambodian project is clearly situated in the centre of the mHealth priorities and concerns in attempting to address issues of sexual and reproductive health amongst young people. The complication is the educational dimension and the cross-over from information-giving and awareness-raising to learning and knowing. Here the mHealth community often seems worryingly naïve and content to pump information through the infrastructure in a way that may be aligned to some models of learning and some cultures of education but not all and not necessarily that of Cambodian young people, who might be exposed to several different versions of learning, from within their families, from their peer group and from the formal institutions such as school.

Mobiles-for-health is only one component of the m4d community and in common with this wider community, it is educationally naïve. These limitations might easily pass unnoticed.

The Discourses of m-learning

Learning with mobile devices, often called *mobile learning* and sometimes *m-learning* is a small but rapidly growing research, development and practitioner community with perhaps 10 years of history, already splintering into different discourses characterising different regional ideas about pedagogy, both formal and institutional, and informal and communal, and reflecting different infrastructural, policy and economic environments.

Much mobile learning in environments where the infrastructure and environment are impoverished has used SMS and so the project is well placed to understand and contribute to the discourses of the mobile learning community. What is, however, less researched and less documented amongst the mobile learning research community and also amongst the distance learning practitioner community is the significance and impact of different conceptions of pedagogy, of teaching and learning, of acquiring knowledge and information in different cultures across Asia and also the differences between these conceptions in formal, official and institutional contexts, those of ministries, schools, exams and text books, and in vernacular, informal and indigenous contexts, those of families, communities and peer groups. Equally under-researched is the significance of the provenance of knowledge and learning in different cultures and sub-cultures, from who are they acquired, with what warranty and for what reasons.

The Cambodian team was based in a formal institutional context, that of a university IT centre, but trying to operate in informal contexts, those of young people, possibly poorly educated, in a subject domain, namely sexual behaviour where hearsay, folklore, gossip and wishful thinking may be the principal forms of knowledge formation and transmission.

A major potential contribution of the Cambodian team was their use of Khmer but this opens up yet more discourses, those of language and linguistics (There is a resonance with South African initiatives at the Meraka Institute[1] to preserve and revitalise indigenous languages such as Xhosa and Tshwane by implementing games and dictionaries online in them).

Mobile learning researchers are just beginning to notice that notions such as personalisation, social constructivism, discursive learning and games-based-learning do not necessarily transfer easily to cultures where individualism, competition, risk-taking or academic status have unexpected connotations.

The Discourses of Software Development

The team was based in the University IT centre and the discourses of the software development community were presumably their mother tongue. These discourses, expressed in the established tomes of software engineering, often focused understandably on the technical product and the technical outcome. Whilst modern variants might emphasise and implement the involvement of the user community, older practices might engage with users only in an initial phase, to *capture the requirement* and in the closing phases to confirm that this requirement had been met, by *user testing*. With a clearly defined, accessible, articulate, knowledgeable and visible user community in a socially and technically stable environment, this may in the past have produced satisfactory outcomes. With an ill-defined and inaccessible community, in this case young people with possible queries about sexual health, working with the rapidly changing social context of a rapidly changing technology, such an approach is problematic. Software engineering practitioners might recommend participative and user-centred development practices, rapid prototyping and iterative and cyclic life cycle models in such cases but implicit in these recommendations is a shift away from tools, rules and prescriptions and towards expertise and judgement. Working in these ways would be desirable, probably necessary, but also challenging, probably intimidating, and would represent a major issue in terms of resources and priorities.

The Discourses of the 'Emergent Researcher'

Furthermore, the concept of an *emergent researcher* itself is not as simple as it sounds and is also highly contextualised. In the context of a mature research institution or a mature research group, an emergent researcher might be a post-doctoral fellow following a clearly defined trajectory of career development and scaffolded progress with local and more senior colleagues. This development might include progressing from second author to first author, from low impact journals to high impact journals, from regional conferences to international conferences, from supporting projects to leading projects, from research associate to principal investigator, and crucially writing funding applications and thereby establishing independence and autonomy. In the context of an institution or indeed a country without an established academic culture, and without routine links into global academic communities, the emergent researcher might be working in isolation with few of the reference points, professional milestones or *rites de passage*, and without access to the various research paradigms, practices and resources, nor the folk-lore and pecking order of conferences, journals, impact factors and peer reviews. They also probably lack the appropriate higher order project management skills, such as scheduling, resource allocation, ethics procedures, and risk analysis and quality assurance necessary for successful multi-person projects with multiple objectives.

We should add that the role and perspectives of the experienced researchers acting as mentors cannot be separated from any account of the teams of emergent researchers with whom they are working. There is a dynamic. The mentors are also the buffer and the conduit between the teams and the funder, adding extra dimensions to their role. These mentors may have broad and eclectic backgrounds but will also inevitably have their own partial and subjective positions. These may include a less nuanced account of the *East/West* context than the *North/South* one. Each will make their own analysis of what is achievable; will strike their own balance between being directive and being supportive and whether their focus should be in the subject domain, in specific research skills or in research project management. In a similar vein, each mentor will have their own experiences and conceptions of research supervision and capacity building (though they may not think these are relevant) leading them to rather different ideas about collaboration and involvement with their team, specifically authoring papers or bidding for funds.

There may also be a tension between aspects of mentoring that relate to contractual compliance, that is monitoring milestones, deliverables, resources and risks, on the one hand, and aspects of mentoring that relate to something collegial and pastoral support, on the other hand. There is also a tension between competing standards, in assessing achievement amongst projects, on the one hand, those standards of 'objectively verifiable indicators' in terms of papers, awards and presentations, and on the other hand, 'value-added' and 'soft outcomes and distance travelled' by emergent researchers.

CONCLUSION

This article has attempted to show that projects, such as that of the team at the Royal University of Phnom Penh, projects not based within a robust and mature research culture but nevertheless attempting to build research potential and to foster emergent researchers, are deeply problematic. They are at the mercy of contending and incommensurable discourses, paradigms and priorities; in terms of *added-value*, the team and others like them make enormous progress but the pressure or ambitions to get recognition as researchers requires not just progress but progress towards and in the terms of a specific research community, and its expectations and priorities. This chapter will not have done justice to the various possible discourses; the point is that such discourses exist.

Finally, these arguments and perspectives have been developed specifically out of the experience of working with the team in Phnom Penh. They must however apply to many other projects.

Notes

1. "Information and Communications" In *Council for Scientific and Industrial Research*, last modified 27 September 2011. <http://www.csir.co.za/meraka/cross_cutting/>.

14

THE DYNAMICS AND CHALLENGES OF ACADEMIC INTERNET USE AMONGST CAMBODIAN UNIVERSITY STUDENTS

Chivoin Peou, May O. Lwin and Santosh Vijaykumar

ABSTRACT

Within the Cambodian educational sector, ICT discourse has brought about new challenges and opportunities. The policy response from the government has been aimed at increasing access to education through ICTs, promoting a life-long learning approach to education, and empowering the Cambodian workforce with technical skills required to compete in a knowledge-based economy, whilst higher educational institutions have promoted the access to the Internet for their students. However, empirical knowledge of their use and understanding of the Internet has been largely absent. This chapter reports the findings from a research project intended to examine the extent to which students at selected Cambodian universities utilise the Internet for general and academic purposes by analysing the dynamics that shape Internet use. At the end, we employ Strengths, Weaknesses, Opportunities, and Threats

(SWOT) analysis to evaluate the current conditions of academic Internet use in Cambodian higher education.

BACKGROUND

The role of Information and Communication Technologies (ICTs) in the upheaval of Cambodia's troubled education sector after the country's emergence from years of political instability has been a topic of much intrigue. Although the quality of and access to education has suffered from a lack of trained teachers and inadequate teaching materials (Abdon, Ninomiya and Rabb 2007; UNESCO 2004), by 2010, the net enrolment rate for primary education reached 95% and that for upper secondary level, 16% (MoEYS 2011; ODI 2010). Additionally, the enrolment rate for tertiary education reached 10% in 2009 (World Bank 2011). These developments were made possible because of the political solution in the early 1990s, particularly the 1991 Paris Peace Agreement and the national reconstruction afterward, which lifted Cambodia out of two decades of isolation and opened up foreign developmental interventions. The ICT for development (ICT4D) discourse in Cambodia is linked to these political changes and developmental interventions, noticeably through the initiatives supported by such international organisations as UNESCO, UNICEF, the Asian Development Bank, and the World Bank (Richardson 2008).

Within the educational sector, the ICT discourse has brought about new challenges and opportunities. In response, in 2004, the Ministry of Education, Youth and Sports (MoEYS) adopted the country's first ICT in education policy entitled *Policy and Strategies: Information and Communication Technology in Education in Cambodia* (MoEYS 2004), followed by its *Master Plan for ICT in Education* in 2009. The policy and the subsequent master plan aim at increasing access to education through ICTs, promoting a life-long learning approach to education, and empowering the Cambodian workforce with technical skills required to compete in a knowledge-based economy. The intended coverage of the policy and master plan ranges from Internet provision and ICT curriculum in secondary school to e-learning in higher education. The Internet is therefore at the centre of the ICT4D discourse in Cambodia with governmental, non-governmental and private institutions promoting its uptake. Whilst affirming voices have encouraged this trend by lauding the Internet's emancipating potential for the developing world (Servaes 2007), dissenters emphasise its potential for catalysing cultural and political domination (Albirini 2008; Servaes 2007). Such polarised arguments and contrasting evidence call for critical empirical inquiries on the advancement of the Internet in Cambodia, especially in education, which is considered a key sector for development.

This chapter explores the extent to which students at selected Cambodian universities utilise the Internet for general and academic purposes by analysing the dynamics that shape Internet use. On the one hand, an understanding of the actual use of the Internet by university students will inform and, potentially, adjust the current enthusiasm toward the integral role of the Internet in higher education, especially in developing countries. On the other hand, empirical data of actual use and the associated dynamics of use are indispensable for critically informing policy making processes. Situating such an inquiry in universities enables us to inform the design and implementation of Internet-based programmes in higher educational institutions in Cambodia.

The study seeks to answer the following research questions:

1. Where do Cambodian university students access the Internet?
2. What is their level of satisfaction at different Internet access points?
3. To what extent do they access the Internet at their universities?
4. What kind of online facilities do they use for academic purposes, and to what extent?
5. What are the main challenges they encounter whilst using the Internet at their universities?
6. What are the perceived benefits of using the Internet for academic purposes?
7. What are the perceived harms of using the Internet for academic purposes?

RESEARCH METHOD

Data Collection

The study commenced with a focus group discussion among university students in the capital city of Phnom Penh that allowed us to generate key themes of interest for the quantitative survey. Subsequently, the research team administered a structured survey questionnaire among students at four higher educational institutions in Phnom Penh. This convenience sampling strategy was adopted as these settings represented a typical university in the city in terms of their affordability (tuition fees charged), academic offering (range of disciplines covered) and provision of Internet access (in computer labs, libraries and in the hallway). Our survey instrument comprised four main sections: socio-demographic factors, Internet-related behaviours and experiences, Internet-related attitudes, and academic utilisation of the Internet.

Academic use of the Internet, motivations for using the Internet, and attitudes towards the Internet were operationalised and measured as follow:

Measures

Academic Internet use was measured by the respondents' responses to the frequencies of six activities on the Internet during the previous six months. The six activities were: (1) Using search engines for school work; (2) Searching websites and archives of government or non-governmental bodies for school work; (3) Looking for news for school work; (4) Downloading books; (5) Communicating with classmates about school work; and (6) Communicating with teachers about school work (Cronbach's α=.84). We created original scales to measure satisfaction with Internet use, perceived challenges of academic Internet use and perceived benefits and harms of academic Internet use.

ANALYSIS AND FINDINGS

We performed uni-variate statistical procedures such as simple frequencies to arrive at broad descriptions of the dataset. Subsequently we performed bi-variate statistical tests such as t-tests and chi-squares to enable a gender-based comparison of means and frequency distributions respectively across key variables of interest.

Sample Characteristics

A broad description of our sample is presented above in Table 14.1. We surveyed a total of 1,565 students, 55% of whom were males and 45% females. The average age of the sample was 21.6 years. Student respondents were derived from more than 30 undergraduate majors categorised into six broad disciplines: arts and humanities (39%), social sciences (22%), business and related (21%), sciences (13%), and law (5%). 80%of the respondents were able to access a PC at home.

The survey attempted to assess the English language skills of participants by seeking self-report scores on their perceived proficiency of the language. Half of the survey respondents reported limited proficiency in English implying that they were able to grasp the general meaning along with some details whilst reading books or news. Nearly a quarter (23%) reported basic proficiency and were thus able to understand the general meaning of only easy text. Nearly a fifth (21%) of the sample reported competent proficiency meaning that they could understand a lot of details and some complex structures/expressions when reading books or news.

TABLE 14.1
Description of Survey Respondents

Sample Characteristics	N = 1565
Gender:	
Male	55%
Female	45%
Education	
Freshman	33%
Sophomore	27%
Junior	22%
Senior	18%
Educational Major:	
Arts & Humanities	39%
Social Sciences	22%
Business-related	21%
Natural Sciences	13%
Law	5%
English-language Competency:	
None	<1%
Incompetent	4%
Basic	23%
Limited	50%
Competent	21%
Good	2%
PC Access at Home:	
Yes	80%
No	20%

Internet Use among University Students

Internet Use Dynamics (Table 14.2): Although a large majority of the students reported having used the Internet, the gender divide was visible through a significant difference between males and females (94% versus 85%, $\chi^2=40.13$, $p<.001$). The majority of the students (63%) had used the Internet for two years or less, whilst only 13% had used the Internet for more than four years, implying that few students came to have experience using the Internet before starting university. In terms of self-perceived Internet ability, nearly three-fourths of the students reported their ability to use the Internet as either poor or fairly competent, whilst only a minority (3%) rated themselves as very competent or expert, who tended to be male. In terms of monthly

TABLE 14.2
Internet Profile of Cambodian University Students

	Total N=1565	Male n=861	Female N=725	χ^2	df	Sig.
Internet use experience						
Yes	90%	94%	85%	40.13	1	.000***
Length of Internet use experience				2.61	3	.455
Beginner (<1 year)	31%	32%	29%			
New (1–2 years)	32%	33%	32%			
Experienced (3–4 years)	24%	23%	25%			
Veteran (>4 years)	13%	12%	15%			
Self-perceived Internet ability				11.82	3	.008**
Poor	25%	25%	23%			
Fairly competent	52%	49%	57%			
Very competent	21%	23%	18%			
Expert	3%	3%	2%			
Monthly expense on Internet				12.98	5	.034*
None	8%	7%	10%			
Up to US$3	49%	47%	51%			
Up to US$5	26%	28%	22%			
Up to US$10	10%	12%	9%			
Up to US$15	3%	3%	3%			
More than US$15	4%	3%	5%			

***Significant at $p < .001$; **significant at $p < .01$; *significant at $p < .05$

expense on Internet use (excluding home connection fees), three-fourths of the respondents reported spending US$5 or less.

General Internet Access Points (Table 14.3): Internet access was greatest at Internet cafés and least at their homes and restaurants offering Wi-Fi access. More male students accessed the Internet at Internet cafés and on their mobile phones as compared to their female counterparts. Although Internet access at home was relatively less, participants reported the highest levels of satisfaction at this venue. Female students reported higher satisfaction with Internet access at home and on mobile phones. Of particular insight is the

TABLE 14.3
General Internet Access Points for Cambodian Students and Related Satisfaction Scores

Use[a] General Access Points	Total N=1552 M (S.D.)	Male n=776 M (S.E.)	Female n=568 M (S.E.)	Female n=692 M (S.E.)	Male n=860 M (S.E.)	Total N=1552 M (S.D.)	Satisfaction[b] General Access Points
School	1.25 (1.42)	1.27 (0.05)	1.23 (0.06)	2.78 (0.05)	2.82 (0.04)	2.72 (0.87)	School
Café***	2.41 (1.30)	2.58 (0.05)	2.18 (0.05)	3.26 (0.03)	3.24 (0.03)	3.30 (0.78)	Cafe
Home	0.86 (1.51)	0.82 (0.05)	0.91 (0.07)	3.35 (0.07)	3.22 (0.07)	3.52 (0.94)	Home**
Restaurant	0.26 (0.76)	0.25 (0.03)	0.28 (0.03)	2.82 (0.08)	2.74 (0.08)	2.92 (0.79)	Restaurant
Mobile*	1.56 (1.71)	1.64 (0.06)	1.44 (0.07)	3.05 (0.06)	2.98 (0.05)	3.14 (0.97)	Mobile*
Others	0.81 (1.27)	0.87 (0.06)	0.72 (0.06)	3.13 (0.06)	3.13 (0.06)	0.77 (0.62)	Others

a: 1=Never to 5 = More than once a week; b: 1=Very unsatisfactory to 5=Very satisfactory
***Significant at $p < .001$; **significant at $p < .01$; *significant at $p < .05$

fact that Internet access provided by the universities was utilised less than that at Internet cafés by more than half, and the universities' provision was even evaluated as dissatisfying by their intended patrons, the students.

University Internet Access (Table 14.4): Within universities we found that participants most frequently accessed the Internet at the Internet lab (63.8%) or in the library (26.9%). Their dissatisfaction towards this university-provided facility was generally attributed to slow Internet speed (54%), limited opening hours (31%), and few available computers (31%).

Further we asked students about different online facilities used for academic purposes (Table 14.5). On a scale of one to five, respondents most frequently used search engines and search directories (M=2.54, S.D=1.46), followed by visiting websites of governmental or nongovernmental organisations (M=1.66, S.D=1.41). Students reported the least use of the Internet to communicate with

TABLE 14.4
Internet Access Points Within University Campuses and
Main Challenges of Internet Access in Universities

Access Point	Total N=1534	Male n=849	Female n=685	Chi	df	P
In University						
Classroom	3.1%	4.2%	3.6%	8.86	1	0.00
Library	26.9%	29.7%	23.5%	7.36	1	0.01
Internet Lab	63.8%	60.8%	67.4%	7.29	1	0.01
Hallway	14.9%	15.5%	14.2%	0.57	1	0.45
None	8.7%	11.0%	5.8%	12.52	1	0.00
Main Challenges						
Cost*	16.7	18.8	14.1	6.12	1	0.01
Limited hours	31.0	30.9	31.2	0.02	1	0.88
Unreliable connection	3.9	4.9	2.8	4.38	1	0.04
Low speed	54.2	54.9	53.4	0.38	1	0.54
Language barrier	15.6	14.2	17.3	2.78	1	0.09
Few computers	31.1	32.2	29.8	1.03	1	0.31
Little assistance	21.0	21.6	20.2			
Lack of understanding	11.8	10.6	13.3	2.70	1	0.10

***Significant at $p < .001$; **significant at $p < .01$; *significant at $p < .05$

teachers about schoolwork or assignments (M=0.77, S.D=1.15) such as use of search engines, websites of governmental/nongovernmental organisations, national and international news and downloading.

Perceived Benefits and Harms of Academic Internet Use (Table 14.6): On a five-point Likert scale (1=Strongly Disagree, 5=Strongly Agree), we found strongest agreement regarding the ability of the Internet to be a good source of learning (M=4.16, SD=0.67) and as an *accessible* source of information for homework (M=4.19, SD=0.70). We also noticed greater than average agreement regarding perceptions of the Internet as a *cheap* source of information for homework, a relatively more convenient information resource than a library and as an enabler of communication between students and teachers. When asked about the potential harms of the Internet, the greatest concerns aroused related to the potential of the Internet to discriminate against those not proficient in English (M=3.42, SD=0.04) and its potential to discourage students from using libraries (M=2.93, SD=0.04).

TABLE 14.5
Main Online Activities Directed Towards
Academic Utilization of the Internet

Online Activities for Academic Work[a]	Total N = 1543 M (S.D)	Male n = 853 M (S.E)	Female n = 690 M (S.E)
Use search engines/search directories[***]	2.54 (1.46)	2.71 (0.05)	2.33 (1.06)
Searching websites of governmental, and non-governmental agencies or organisations[***]	1.31 (1.30)	1.45 (0.04)	1.14 (0.05)
Search for current national and international news[***]	1.66 (1.41)	1.86 (0.05)	1.40 (0.05)
Download books[***]	1.23 (1.39)	1.44 (0.05)	0.98 (0.05)
Communicate with classmates about school work or assignments	1.25 (1.34)	1.27 (0.04)	1.22 (0.05)
Communicate with teachers about school work	0.77 (1.15)	0.81 (0.04)	0.72 (0.04)

a: 1=Never to 5=More than once a week
[***]Significant at p < .001; [**]significant at p < .01; [*]significant at p < .05

DISCUSSION

As the number of Cambodian Internet users has rapidly grown during the last decade, from 2,000 in 1998 to 74,000 in 2008 (ITU 2011) and to about 290,000 in 2010 (according to the Ministry of Post and Telecommunication's estimate [DMC 2010]), it is likely that the Internet will become more integrated into the lives of youth in the relatively advantaged socioeconomic bracket, including university students. Therefore it becomes increasingly important to understand their preferences and opinions about the existing state of ICTs in Cambodia's universities before strategising such investments that directly influence them. We present a discussion of our findings in the form of Strengths, Weaknesses, Opportunities, and Threats (SWOT) analysis. Our approach is consistent with Peizer's (2003) postulation that planners and leaders who incorporate ICTs into projects in less developed countries must work "within the constructs of the existing system by leveraging its strengths

TABLE 14.6
Perceived Benefits and Harms of Academic Internet Use

Perceived Benefits[a]	Total N=1552	Male n=860	Female n=692	Female n=692	Male n=860	Total N=1552	Perceived Harms[a]
Accessible source of information for homework	4.19 (0.70)	4.20 (0.02)	4.17 (0.03)	3.42 (0.04)	3.27 (0.04)	3.33 (1.07)	Internet discriminates those not proficient in second language[*]
Cheap source of information for homework	3.44 (0.98)	3.46 (0.03)	3.42 (0.04)	2.82 (0.04)	2.90 (0.03)	2.86 (0.98)	Expensive source of information for homework
Internet more convenient than a library	3.69 (0.87)	3.67 (0.03)	3.71 (0.03)	2.93 (0.04)	2.81 (0.03)	2.86 (0.98)	Because of Internet students avoid using libraries[*]
Enables communication between students and teachers	3.52 (0.92)	3.53 (0.03)	3.51 (0.04)	2.70 (0.03)	2.71 (0.03)	2.70 (0.88)	Online communication with teachers about school work is inconvenient
Internet is a good source for learning[*]	4.16 (0.67)	4.19 (0.02)	4.11 (0.03)	2.82 (0.04)	2.80 (0.03)	2.81 (0.97)	Internet full of unreliable, misleading information

a: 1=Strongly Disagree to 5=Strongly Agree
***Significant at p < .001; **significant at p < .01; *significant at p < .05

and limiting its weaknesses" (p. 87). We believe that such presentation enables policymakers and programmers to optimise the benefits of academic inquiry and catalyses a swift translation of empirical research into policy priorities.

Strengths. The Internet is increasingly integrated into the daily lives of Cambodian youth as evidenced by our study findings, which show that nearly 90% of students had accessed the Internet. Whilst the Internet was still mainly used for information seeking, it was evident that it was also utilised for entertainment and socialisation purposes, and the students generally felt positive toward the development of the Internet. Such integration is expected to gain additional momentum with an increased maturation and accessibility of the technology. The increase in the accessibility of the Internet is attributable to two main factors: the rising number of Internet Service Providers (ISPs which reached 34 in 2009, according to data from the Ministry of Post and Telecommunication [DMC 2010]) and consequently, a wider range of access points such as mobile phone and wireless modem.

Among the various Internet points, we studied Internet cafés, locally known as 'Internet shops', which remain the most crucial access point for users. In the developing Cambodian economy, Internet cafés might become indispensable as they enable users to avoid infrastructural and hardware set-up costs. In such a context, Internet cafés can actually be harnessed for educational potential, particularly given their informal and unrestricted environment for autonomous, self-directed and mature learners (Cilesiz 2009).

Another implicit strength that has emerged from this research is that Cambodian youth seem intent on seeking Internet access despite low penetration rates from a range of access points including mobile phones. Whilst intention to seek Internet access remains to be captured and measured as a theoretical construct in future research, the patterns of Internet use demonstrated by study participants demonstrates that future investments in ICT for education initiatives in Cambodia might be ensured of steady technological uptake by their intended recipients.

Weaknesses. Our study showed that provision of the Internet by Cambodian universities is far from adequate. Students seemed less satisfied with Internet access and online facilities because of slow speed, limited accessibility hours, few computers and minimal assistance. In relative terms, their dissatisfaction about Internet access in universities was stronger as compared to Internet access at home or in Internet cafés. It is also apparent by the minimal level of online communication between students and teachers that the Internet has yet to be strategically integrated into systems of teaching and learning. Indeed, any rationale other than educational should not dictate an investment of Internet provision or integration into higher education (Broad, Matthews and McDonald 2004).

Opportunities. Participants ranked mobile phones as an Internet access point more than university-based sites and lesser than Internet cafés. The trend was similar between both genders. Mobile phone penetration has substantially increased in Cambodia between 2005 (just over one million subscriptions) and 2010 (over eight million subscriptions) (ITU 2011). Whilst the penetration rate remains low in comparison to other countries in the Asian region, and service sometimes unreliable, Cambodia's cell phone market is massive in comparison to its Internet service. The quicker diffusion of mobile phones as opposed to PC-based Internet provides immense potential for educational institutions to generate innovative mobile media for ICT for Education initiatives. Utilisation of cell phone modules and applications for education are already commanding attention in developing economies such as India and South Africa and such a direction might provide cost-effective solutions in the long term. In order to catalyse this leapfrogging behaviour it will be necessary to address infrastructural impediments to efficient cell phone delivery in the short and medium term — a necessary step to ensure mobile-based Internet access for Cambodian students.

We found that students strongly perceived the benefits of Internet access for academic use in terms of facilitating their homework assignments and as an information source for learning. Whilst existing Internet access facilities have thus far been utilised by survey respondents for generic information seeking such as using search engines and for news, findings about the perceived benefits point to future directions in which students might use the Internet. These findings potentially inform new opportunities in terms of educational strategies that can be used by educators who are integrated into ICT4D efforts.

Threats. The utilisation of the Internet for academic purposes remains marginal by students, highlighting an underutilisation, or even inadequate understanding, of the Internet and its potential by both the learners and educators. Also worth noting is that although half of the students used search engines at least once a week to get information for their academic work, this could be problematic due to wide variations in the quality of online information. As training programmes on Internet search and related skills do not exist in school and university curricula, there are associated risks for using search engines for academic work: encountering overwhelmingly abundant information on the one hand, and consuming unreliable or misleading information, on the other. These findings bear two implications. One, students need to be aware of the immense resources available at no cost (such as open courseware, *Slideshare* and *YouTube* tutorials). Concurrently, it is important for educational institutions to enhance their awareness with regards to issues such as Internet privacy, plagiarism, and strategies to discern the reliability of online information sources. These measures will likely expand their utilisation of the Internet for educational purposes beyond search engines and enhance professional and educational opportunities in the long run.

Another threat pertains to Internet skills and English-language proficiency of Cambodian students. Whilst 52% of the respondents in our sample rated themselves as fairly competent in Internet skills, 50% self-reported their English-language proficiency as limited. This is a cause for concern as there was a relatively strong concern among the respondents that the Internet could discriminate those not conversant in a second language (i.e. English). The findings imply that moving forward ICT for Education efforts in Cambodia might confront the critical challenge of Khmer language-based online facilities in order to encourage greater uptake, use and diffusion.

Lastly, significant gender differences in the extent of use of online activities for academic purposes reveal the presence of a gender-based digital divide. Male students were found to demonstrate consistently higher engagement in such activities, especially the use of search engines, visiting governmental/ non-governmental websites and downloading books, when compared to their female counterparts. An explanation for this finding possibly lies in the stronger perception among female students about the Internet as a force that can (a) discriminate those with lower second-language proficiency; and (b) lead students away from traditional libraries. Our findings are broadly consistent with cyber-psychology literature about more negative attitudes among females with regards to computers and the Internet (Broos 2005). Policy-makers may prioritise this issue moving forward lest the infusion of ICTs in higher education threatens to perpetuate a gender bias in accrued benefits and potentially exacerbate a knowledge divide among Cambodian youth in the long term.

CONCLUSION

In this study we surveyed over 1,500 students at four higher educational institutions in Phnom Penh on their Internet use for academic purpose by analysing the dynamics that shape their Internet experience. On a positive note, we found that a large majority of them (90%) had experience using the Internet; the Internet cafés, as an inexpensive public access venue, offered a satisfying experience of accessing the Internet, thus offsetting the relatively high cost of owning the hardware to use the Internet at home, and were optimistic about the Internet as a good source of learning and an accessible source of information for homework. In addition, as mobile subscription has seen an exponential growth over the last few years together with growing availability of mobile phone Internet and wireless modem, there is a strong potential of harnessing mobile technologies for education initiatives. One of the main disappointments is, however, the fact that Internet access at university facilities was found dissatisfying by the students, rendering thorough attention to improving the service. There are also threats that demand attention, which include the underutilisation of the Internet for improving learning experiences among the students and the gender-based digital divide in utilising the Internet for academic purpose.

References

Abdon, Buenafe R., Seishi Ninomiya, and Robert T. Raab." e-Learning in Higher Education Makes Its Debut in Cambodia: The Provincial Business Education Project." *International Review of Research in Open and Distance Learning* 8 (2007): 1–14.

Albirini, Abdulkafi. "The Internet in developing countries: a medium of economic, cultural and political domination." *International Journal of Education and Development Using ICT* 4 (2008).

Broad, Martin, Martin Matthews and Andrew McDonald. "Accounting education through an online-supported virtual learning environment." *Active Learning in Higher Education* 5 (2004): 135–51.

Broos, Agnetha. "Gender and Information and Communication Technologies (ICT) Anxiety: Male Self-Assurance and Female Hesitation." *CyberPsychology & Behaviour* 8 (2005): 21–31.

Cilesiz, S. "Educational computer use in leisure contexts: a phenomenological study of adolescents' experiences at Internet cafés." *American Educational Research Journal* 46 (2009): 232–74.

DMC. *Cambodian Communication Review 2010*. Phnom Penh: Department of Media and Communication, December 2010.

ITU. *Key 2000–2010 Country Data*, 2011. <http://www.itu.int/ITU-D/ict/statistics> (16 November 2011).

MoEYS. "Policy and strategies on information and communication technology in education in Cambodia," 2004. <http://www.moeys.gov.kh/DownLoads/Publications/ict_edu_en.pdf> (1 May 2011).

———. "*Summary report on the education, youth and sport performance in the academic year 2009–2010 and the academic year 2010–2011 goals*," March 2011. <http://www.moeys.gov.kh/en/component/content/article/135-education-congress-report/125-summary-report-of-education-congress-2009-2010.html> (1 May 2011).

ODI. "Cambodia's story: Rebuilding basic education in Cambodia: establishing a more effective development partnership," 2010. <http://www.developmentprogress.org/sites/default/files/cambodia_education_0.pdf> (1 May 2011).

Peizer, Jonathan." Cross-sector information and communication technology funding for development: What works, what does not, and why." *Information Technologies and International Development* 1 (2004): 81–88.

Richardson, Jayson W. "ICT in Education Reform in Cambodia: Problems, Politics, and Policies Impacting Implementation." *Information Technologies and International Development* 4 (2008): 67–82.

Servaes, Jan. "Communication for development: making a difference." Paper presented at the The World Congress on Communication for Development: lessons, challenges, and the way forward, 2007. Rome: FAO and WB.

UNESCO. *Metasurvey on the use of technologies in education in Asia and the Pacific on the UNESCO website*, 2004. <http://www.unescobkk.org/fileadmin/user_upload/ict/Metasurvey/COMPLETE.PDF> (1 May 2011).

World Bank. 2011. <http://data.worldbank.org/country/cambodia> (16 November 2011).

SECTION IV

Synthesis and Conclusion

15

FINDING A PATH TO
INFLUENCING POLICY

Roger Harris and Arul Chib

The path from research to practice, or vice versa, is neither easy to find nor to follow. Having claimed at the outset that academics and practitioners rarely associate with each other, the output from the research that has been described in this book suggests that there is considerable potential for desirable outcomes if they did. Whilst it is relatively straightforward to identify the **potential** for practical impact from the SIRCA research projects, it is not so clear-cut when trying to isolate **actual** influences and even less so when trying to measure them. As a starting point it is probably useful to agree on a typology of research impacts and it is possible to represent them in Figure 15.1.

Academic impact is achieved by publishing papers in scientific journals and then having them cited by other researchers. The publication process ensures that a body of knowledge is created in a central repository; whilst citations are then attempts to build upon that body of knowledge. Knowledge is thus created incrementally, with existing knowledge subject to testing, validation, and criticism. Simultaneously, the centralised knowledge then becomes a source for the training and development of future scientists, who in turn add to the body of knowledge. Whilst this process theoretically exists as a means to continuous improvement, in practice the process often devolves to mere number-counting of publication and citations as indication of quality.

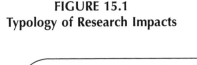

FIGURE 15.1
Typology of Research Impacts

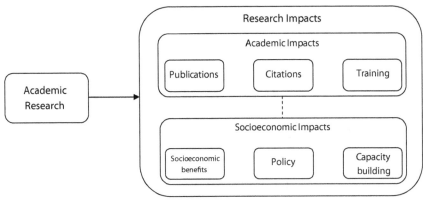

   Publications and citations are counted, categorised, and ranked; the
more a researcher receives, the better his/her prospects for promotion, tenure
and continued employment in the long run, and in the short run, bonuses,
grants, and titles are bequeathed on this basis. Increasingly, scientific journals
are ranked according to their perceived prestige, and the higher the rank
of journal, the more a paper published in it counts towards professional
development. A report by a UK parliamentary select committee on peer
review concluded, "Representatives of research institutions have suggested
that publication in a high-impact journal is still an important consideration
when assessing individuals for career progression," (2011). In the better
universities (determined once again by multiple ranking systems), scores are
kept of academic publications and citations, with points assigned according to
the journal ranking. No other form of publication or research dissemination
contributes towards academic impact. In this publish-or-perish environment,
other forms of research impact receive scant attention, and the reader will
notice in the diagram above that there is a limited relationship between
academic impact and socioeconomic impact. Whilst this might seem a
pessimistic approach, we consider describing the limited impact of research
on socioeconomic indicators as rather optimistic, given the state of most
research conducted within the ivory towers of universities.

The second form of impact; socioeconomic, is depicted as consisting of three components. In the first place, the research may have direct socioeconomic benefit for those who participated. The villagers in Vietnam, described in the chapter *Integrating Digital and Human Data Sources for Environmental Planning and Climate Change Adaptation: From Research to Practice in Central Vietnam*, clearly benefited from learning better how to adapt their agricultural practices to climate and climate change related events. The second type of socioeconomic research impact comes from its influence on policy formation, which would include contributions that steer public debate around key social issues. The lessons learned from the Philippine study on political blogging, described in the chapter *It's the Talk, not the Tech: What Government Should Know About Blogging and Social Media,* point directly to the nature of the relationship government has with its citizens and how this can be mediated in cyberspace towards greater democratic engagement. Finally, research can build capacity in its participants and this has the potential for further impact. In the chapter on *Messy Methods for ICT4D Research*, we learned about the inherent "messiness" of ICT4D research, which, although not discussed there, has its counterpart in practice. In fact, the socio-technical approach to problem solutions that involve ICTs claims that they are inherently "messy" in that as soon as one begins to address a problem with ICTs, the nature of the problem itself changes. Just like the researchers, implementers and participants need to proceed in a process of mutual learning that implies frequent adaptations and adjustments away from the pre-defined course of action, and this becomes problematic for project managers who are restricted within the bounds of a project's logical framework. The chapter, *ICTD Curriculum Development and Professional Training: Mainstreaming SIRCA Research Models* illustrates how the other research projects can contribute to the training needs of practitioners by exposing the range of skills and methods that are required within ICTD interventions that result in desirable outcomes. Also, research in support of policy advocacy and improved practice reinforces the need for ethical oversight, as outlined in the chapter *Ethics and ICTD Research*, because the impacts will potentially affect the lives of people, and their well-being imposes considerations that may not always be relevant when the only output is an academic publication.

In highlighting the absence of a connection between these two groups of impact, it should also be emphasised that achieving either one does not automatically lead to achieving the other. Moreover, there are considerable differences in the processes that lead to each and whilst conducting research that is recognised as being of high quality is always the starting point for both, there is probably a need for at least as much attention to be paid to the achievement of the non-academic impacts as there is to fostering the high-quality research that can have the potential for doing so. The first, and probably most significant, difference is that academic research has to break out of its closed loop mode of operation and adopt less of a knowledge-push model and more of a knowledge-pull approach. What this implies can represent a sizeable break from the customary approaches that researchers take towards their work; involving new relationships and power exchanges within networks of non-linear influences and uncertain outcomes. It is therefore not surprising that few attempt it, but the rewards can be considerable as long as they are recognised and actively pursued at the institutional level.

One set of key stakeholders largely ignored, except in the case of obtaining large-scale grants, are policy-makers at government and transnational levels, as pointed out in the chapter, *Multi-Stakeholder Perspectives Influencing Practice-Research-Policy*. There have been systemic failures in translating the output of academic research to wider national level programmes. However, one might ask, why should academic institutions move out of the closed loop of research into the messy field of policy? One good reason; a question raised in our opening chapter, "What has been the overall impact of ICT4D in agriculture, in education, in health, in governance, in natural resources management?" "Where is the evidence?" Basic research and development within, and evaluation of, ICTD programmes have the potential to provide a solid base of evidence for investment in innovative transformative programmes. Many policy circles are moving towards an evidence-based approach to policy formulation and programme design and are seeking the evidence they need to justify their actions. So the first reason to provide such evidence is that the market for the knowledge that academic research can generate is already well established. Whilst many institutions accept this as a justification for their work, the main challenge they face lies in engaging with that market with mechanisms that will increase the likelihood of their output having some influence. This is important not least because effective policy is required to support and advance innovations, as well as to guide both public opinion and that of research to areas of research that have emergent ethical and moral implications, such as privacy issues (WHO 2011). One depiction of a taxonomy of mechanisms for supporting the flow of knowledge that can be useful in navigating a route into the realm of non-academic impacts is as shown in Figure 15.2 (van Heijst et al. 1998).

The framework is adapted to the context of research communication from its origins in the management literature on organisational memory systems. It has two dimensions; the demand for and supply of information, and whether these demands are active or passive. This results in four spaces in which information exchange occurs, characterised as follows:

FIGURE 15.2
Knowledge Flow Matrix

Knowledge Demand

		Passive / Latent	Actively requesting
Knowledge Supply	Passive. Distribute standard products	The Knowledge Attic	The Knowledge Publisher
	Active. Distribute customised products	The Knowledge Pump	The Knowledge Dialogue

Source: van Heijst et al. (1998).

1. The Knowledge Attic; an archive in which material is collected and stored.
2. The Knowledge Pump; a system which tries to deliver information selectively to people who are believed to need it.
3. The Knowledge Publisher; a system which allows users to actively request information, but which does not respond with a tailored supply of information.
4. The Knowledge Dialogue; a system which enables users to request information and which responds with a corresponding supply.

By mapping academic research onto this framework as the Knowledge Supply and the policy and practice community as the Knowledge Demand, the earlier depiction of academic research as a closed loop would imply that its output constitute a Knowledge Attic, the simplest form of knowledge exchange mechanism; an archive (library) in which material is collected and stored, where there is little interaction with the demand side of the framework. In contrast, in a 2009 study by DFID of 17 research communication programmes it was supporting, totalling an estimated budget of US$11.34 million (£8.7 million), the contribution by value was approximately distributed as follows; 19% to knowledge attics, 18% to knowledge publishers, 52% to knowledge pumps, 11% to knowledge dialogue (DFID 2009). Even though more than 80% of the research investment was categorised within the Knowledge Publisher, Knowledge Pump and Knowledge Dialogue mechanisms, the DFID study concluded with recommendations that emphasise their strengthening and further embedding within wider development processes. This is then suggested as the target for academic institutions that want their products to have socioeconomic impact and relevance.

We have seen in chapter, *ICTD Praxis: Bridging Theory and Practice*, that the route to non-academic impact for academic research is littered with challenges in collaboration that include differing incentives, goals and processes, leading to an inappropriate balance of academic research and practicable solutions, with limited emphasis placed on knowledge translation for reaching users and audiences for the research. We have also seen in the chapter, *From Production... To Dissemination... To Adoption*, how the SIRCA programme evaluations emphasised the importance of disseminating research findings to academic and non-academic audiences by, for example, publishing non-technical content in mass media for dissemination to decision-makers and the general public. There is much to be said for such an approach in terms of widening the audience, with however, always uncertain results that potentially range from insignificant to profound, but it might be regarded as a scatter-gun mechanism with no specific target but a good chance of hitting something. But as we established in the opening chapter, *Perspectives on ICTD Research and Practice*, the path from research to practice is a long one and requires more than dissemination. An additional approach is therefore proposed; one that re-positions academic knowledge exchanges within the framework above and which actively targets policy and practice outcomes from the research. In this approach, the required socioeconomic impact is the starting point, addressed by asking the question, "What is the desired policy and/or practice impact that the research will generate?" Such an approach can be characterised as policy entrepreneurialism (ODI 2009).

Policy entrepreneurs within the context of ICTD are researchers wishing to maximise the impact of their research on policy and programme formulation in order to inspire and inform policy and practice that leads to the reduction of poverty. As policy and practice advocacy becomes the starting point and the intended research impact, it is well to acknowledge some fundamental aspects of policy processes in order to design research that can effectively influence them. Firstly, policy processes are complex and mostly non-linear, non-sequential and non-logical so that the presentation of information alone will rarely affect them. Many empirical studies have shown that only rarely will research impact be direct, instrumental and clearly identifiable, such as when research leads directly to specific policy choices, or when research is neatly captured and codified in tools and instruments such as guidelines, protocols or organisational processes. Instead, much important decision making is diffuse, and characterised by 'non-decisional processes' and the progressive establishment of new routines (Weiss 1982). Many actors are involved and they interact within an intricate network of communication and exchange.

Secondly, processes for policy and programme formulation are only weakly informed by research-based evidence. Policy-makers tend to be influenced more heavily by; their own values, experience, expertise and judgment, the influence of lobbyists and pressure groups; and pragmatism, based on the resources available, than they are by evidence (Davies 2005). However, research may also be absorbed and internalised into professional tacit knowledge as it emulsifies with many other sources of knowledge (experience, anecdote, received wisdom, lay knowledge etc.). In doing so, it may leave few tell-tale signs of its passage, role or impact. Thus, research can contribute not just toward decisional choices, but also toward the formation of values, the creation of new understandings and possibilities, and to the quality of public and professional discourse and debate. Capturing these subtle and diverse impacts poses considerable conceptual, methodological and practical challenges (Davies et al. 2005). Next, policy entrepreneurs need additional skills to influence policy and practice. They need to be political fixers, able to understand the politics and identify the key players. They need to be good storytellers, able to synthesise simple compelling stories from the results of complex research. They also need to be good networkers to work effectively with all the other stakeholders, and they need to be good engineers, building a programme that pulls all of this together. Or they need to work in multidisciplinary teams with others who have these skills. Finally, policy entrepreneurs, like all entrepreneurs, need to be highly motivated and committed.

According to ODI (2009), turning a researcher into a policy entrepreneur, or a research institute or department into a policy-focused think tank is not easy. It involves:

1. *A fundamental re-orientation towards policy engagement rather than academic achievement*; this is a massive challenge. Universities attach enormous importance to the various rankings that appear from time to time as they are instrumental in attracting students and funding. The QS World University Rankings ranks the top 400 universities in the world. Its criteria are dominated (40%) by the universities' academic reputations from a global survey of 15,000 academic respondents, plus citations per faculty (20%). Another 20% of the measurements are derived from the faculty-student ratio. Nowhere is there any reference to policy influence (QS World University Rankings 2011). Accordingly, energies that are directed towards policy advocacy which do not also influence the criteria

used for deriving rankings threaten an institution's ability to survive in its bread-and-butter business.

2. *Engaging much more with the policy community*; by and large, for academics, this is a community of strangers. For the reasons previously stated, academics and the policy and practice community do not share the same space and both are unsure of how to break the ice and begin any form of dialogue. Identifying the user community is less straightforward than researchers might imagine. It has been observed that researchers and research funders have succumbed to the temptation of constructing and then believing in users of their own making. Actual and potential users may not map so readily to those identified and vaunted *a priori* by research teams, research funders or research impact assessors (Shove and Rip 2000).

3. *Developing a research agenda focusing on policy issues rather than academic interests*; this need not present a dilemma as policy issues can easily coincide with academic interests. In the fast-moving and emerging sphere of ICTD policy, there is an abundance of research topics that can easily satisfy the demands of a theoretically-dominated research enquiry. A significant problem in ICTD research has been that academic measures, derived after much painstaking work in laboratories, seminars, and field surveys, rarely map onto those that policymakers require to make decisions. The challenge is for both sets of stakeholders to find common ground; for example, in defining indicators and/or measures that are relevant to both academics and policymakers.

4. *Acquiring new skills or building multidisciplinary teams*; however, this is easier said than done — the ICTD field itself is multi-disciplinary, composed of a diverse set of scientists ranging from those in the hard sciences of computing and information technology, to those in the social sciences of sociology, psychology, and communication. Finding a common lexicon can often be the critical barrier to collaboration, whilst the lack of common literature or theoretical frames hampers immensely (Chib 2009). However, most contemporary ICTD research benefits from multidisciplinary approaches anyway, and we have seen already in previous chapters how this contributes to the gathering of data and the interpretation of findings. This book is an attempt to bring together an array of voices that provide substance to the argument that multidisciplinary endeavours can produce effective outcomes.

5. *Establishing new internal systems and incentives*; the provision of incentives for policy-oriented research has to be capable of compensating for any perceived loss of academic credibility suffered by researchers whose

principal output is not necessarily a publication in a top journal. Internal systems that value policy-orientations in research activity can support those who engage in it and they can legitimise their interests within their professional realm.

6. *Spending much more on communications*; academic research projects typically set aside between five and ten percent of their budgets for dissemination, covering the costs of attending a conference and writing a paper. Policy-oriented research budgets typically provide upwards of 70% of their budget for communication activities, which cover a wide range of actions and outlets within a network of diverse actors.

7. *Producing a different range of outputs*; whilst the typical journal publication can be a natural corollary of policy research, it is not sufficient to achieve policy influence. Other media products are required but again these are unlikely to be sufficient for serious impact. Moreover, the process of journal publishing is fraught with barriers, the first being cost. For example, the average cost of an annual subscription to a chemistry journal is US$3,792. A single article published by one of Wiley-Blackwell's journals will cost US$42. Additionally, it is claimed that far from assisting the dissemination of research, the big publishers impede it, as their long turnaround times can delay the release of findings by a year or more (Monbiot 2011). Finally, the volume of academic publishing, estimated at 1.3 million papers in 23,750 journals in 2006 (Björk et al. 2009) and the imperatives pushing the production of more, result in a huge mass of impenetrable knowledge.

8. *Working more in partnerships and networks.* The policy-practice sphere is far more diverse, open and imperceptible and its shape as a network is a far cry from the hierarchical format with which academics have become accustomed. As influences arrive from a range of sources, those of the academic compete with and have to be blended with others within a rapidly shifting dynamic that is difficult to assess and impossible to measure; all new territory for the average academic researcher who has not engaged before with it.

9. *It may also involve looking at a radically different funding model.* Most academic research is publicly funded, yet donors spend billions annually on development research; an estimated US$2.7 billion each year. DFID doubled its research budget from US$151 million (£116 million) GDP in 2006/07 to US$286 million (£220 million) by 2010/11 (Jones and Young 2007). ODI has reported that DFID and IDRC are the only two international donors to have identified research utilisation and communications as a priority focus area, but it also noted that;

"nevertheless, several key informants highlighted the need to undertake more rigorous evaluations of what types of research-policy linkages and research communication and utilisation strategies are effective in different research fields and political and policy contexts, suggesting that this was an important under-researched area. It was also emphasised that a focus on research into use should not be conflated with embedding policy research questions in research design. In this regard, specific attention to building the capacities of southern policymakers to become more effective and informed consumers of knowledge was identified as an area that had received insufficient attention to date (ibid.)." Our conclusion is that donors are a fruitful source of research funding, as exemplified by the SIRCA programme, and also that there is scope to develop the market further for the knowledge that such research generates, which includes educating the recipients as to what is, and could become, available.

As we report on the SIRCA programme (SIRCA I), deliberations are under way for the next round, SIRCA II, which will carry the original objectives into a wider geographic footprint. The primary goal remains clear and vital; to identify future research leaders and to facilitate their development through the support of research grants, building their capacities to conduct research in ICTD. However, there is an intent within SIRCA II to extend beyond a more-of-the-same approach with an increased emphasis on research take-up; the utilisation of the knowledge generated by research within the fields of policy advocacy and programme practice through the packaging of research in a more suitable manner such that it can be presented to policy-makers and have direct or even indirect impacts on policy.

SIRCA II now presents the opportunity to develop emergent researchers as potential policy entrepreneurs, within institutions such as the Singapore Internet Research Centre at Nanyang Technological University, Singapore, the University of KwaZulu-Natal, South Africa, and Instituto de Estudios Peruanos, Peru, that have the capacity to act as knowledge brokers with an interest in influencing ICTD policy development at the national and international level, wholly in line with objectives of IDRC, the SIRCA programme donor. The second round will achieve the same academic impact in terms of peer-reviewed publications that remain a priority within the professional context, but will add this new layer of socio-economic impact in order to further promote ICTD research as a tool for advocating for policy formulation and enhancing programme practice.

It is irrefutable that there is pressing demand for this. The pace of innovation with ICTs and potentially within ICTD has never been greater.

Information is becoming more liquid globally; mass communication channels are now out of the monopolistic control of governments and corporations; the drive for openness in all areas of development is intensifying. At the same time ITU reports that 5.2 billion people are still not using the Internet and that in the developing world only eight per cent of rural communities have Internet access (ITU 2008). Mobile devices are everywhere except in the poverty statistics, despite the many promising examples of their use for poverty alleviation. The need to address these issues with evidence-based advocacy derived from rigorous and relevant research remains as strong as ever and at the forefront of SIRCA II.

References

Björk, Bo-Christer, Annikki Roos and Mari Lauri. "Scientific journal publishing: yearly volume and open access availability". *Information Research* 14 (2009).

Chib, Arul. "The Role of ICA in Nurturing the Field of Information and Communication Technologies for Development". *International Communication Association newsletter* 37 (2009): 13–16.

Davies, Huw, Sandra Nutley and Isabel Walter. "Assessing the impact of social science research: conceptual, methodological and practical issues". ESRC Symposium, May 2005.

Davies, Phil. "Evidence-Based Policy at the Cabinet Office". Impact to Insight Meeting, ODI, London, October 2005.

DFID. *Learning lessons on research uptake and use: A review of DFID's research communication programmes.* 12 June 2009.

International Telecommunication Union (ITU). "Global ICT Developments," last modified 15 July 2008. <http://www.itu.int/ITU-D/ict/statistics/ict/index.html>.

Jones, Nicola, and John Young. "Setting the scene: Situating DFIDs Research Funding Policy and Practice in an international Comparative Perspective". Draft paper, 2007. London: ODI.

Monbiot, George. "Academic publishers make Murdoch look like a socialist". *Guardian,* 29 August 2011. <http://www.guardian.co.uk/commentisfree/2011/aug/29/academic-publishers-murdoch-socialist>.

ODI. *Helping researchers become policy entrepreneurs How to develop engagement strategies for evidence-based policy-making.* Briefing Paper 53, September 2009.

"QS World University Rankings". 2011. <http://iu.qs.com/projects-and-services/world-university-rankings/>.

Shove, Elizabeth and Arie Rip. "Users and unicorns: a discussion of mythical beasts in interactive science". *Science and Public Policy* 27 (2000): 175–82.

UK Parliament. "Commons Select Committee Peer Review," 2001. <http://www.parliament.uk/business/committees/committees-a-z/commons-select/science-and-technology-committee/inquiries/peer-review/>.

van Heijst, Gertjan, Rob van der Spek and Eelco Kruizinga. "The lessons learned cycle". In *Information Technology for Knowledge Management*, edited by Uwe M. Borghoff and Remo Pareschi, pp. 17-34. Berlin: Springer Verlag, 1998.

Weiss, Carol. "Policy research in the context of diffuse decision making". *Journal of Higher Education* 53 (1982): 619–39.

World Health Organisation (WHO). *Global Observatory for eHealth series, Volume 3: mHealth: New horizons for health through mobile technologies.* 2011.

ABOUT THE CONTRIBUTORS

THE EDITORS

Arul Chib, PhD

Assistant professor at Nanyang Technological University, and is the assistant director of the Singapore Internet Research Centre. He examines the impact of development campaigns delivered via a range of innovative information and communication technologies (ICTs), particularly in the arena of mHealth, where his research programme has extended to the use of mobile technology by remote healthcare workers in the rural regions of China, India, Nepal, Thailand, and Uganda. He has contributed to the conceptual progress of the discipline by proposing theoretical frameworks of analysis, including the ICT for healthcare development model, and the Technology-Community-Management model. Dr Chib has published in international refereed publications, such as the *International Journal of Communication*, *Journal of Computer-Mediated Communication*, *New Media and Society*, *Asian Journal of Communication*, *Information Technologies & International Development*, and the *Journal of Health Communication*; and has contributed to various book chapters from leading publishers such as Cambridge Scholars, Hampton Press, Springer, and Taylor & Francis. Dr Chib is the winner of the 2011 Prosper. Net Scopus Young Scientist Award in ICTS for Sustainable Development, awarded by Elsevier and the United Nations University.

Roger W. Harris, PhD

Founder of Roger Harris Associates (RHA), a consulting and social entrepreneurial firm. Its mission is to alleviate poverty by facilitating access to Information and Communications Technologies (ICTs) for underserved sections of society in Asia, and to ensure they are able to make good use of the technology according to their own development aspirations. Dr Harris has been working in this field since 1997 and the firm has been providing consulting and knowledge-sharing services for eight years in the area of ICTs for poverty reduction and rural development in Asia. Services include policy advice and development, programme design and evaluation services, project implementation assistance, and research to governments, international development agencies, and civil society bodies.

THE AUTHORS

Komathi Ale
Ph.D. Candidate at the University of Southern California's Annenberg School for Communication and Journalism. She received both her B.A. and M.A. degrees in Communication Studies from the Wee Kim Wee School of Communication and Information at Nanyang Technological University in Singapore. Her research focuses on the impact of information and communication technologies for development, with specific interest in addressing social issues amongst marginalised communities.
KOMA0004@e.ntu.edu.sg

Rahul De', PhD
Hewlett-Packard Chair Professor in Information and Communication Technology (ICT) for Sustainable Economic Development, Quantitative Methods and Information Systems Area at the Indian Institute of Management in Bangalore.
rahul@iimb.ernet.in

Laurent Elder
Laurent Elder leads IDRC's Information and Networks programme, which aims to better understand how information networks can be used to promote open, inclusive and rights based information societies in the South. From 1999 to 2004, Elder was in Senegal at IDRC's office for West and Central Africa. There, he worked with the Acacia initiative, a programme looking into the impact of information and communication technologies in Africa. Elder has also worked at Canada's Department of Finance and began his career in France, researching issues related to French history and international relations.
lelder@idrc.ca

Alexander Flor, PhD
Professor of Information and Communication Studies at the University of the Philippines - Open University. Formerly UPOU Vice Chancellor for research and development, he was the founding Dean of the Faculty of Information and Communication Studies serving two terms (2004 to 2010).

Professor Flor was the first Knowledge Management (KM) Program Manager (1998-2002) of SEAMEO SEARCA, a regional organisation based in Los Baños. In 1998, he designed, developed and introduced KM as an academic course under the UP Los Baños development communication programme.
aflor@upou.edu.ph

Ma. Regina M. Hechanova, PhD
Executive Director of the Ateneo Centre for Organisation Research and Development, Ateneo De Manila University (AdMU) in the Philippines. She is a faculty of the Industrial/Organisational (I/O) Psychology at the AdMU. A prolific researcher, she was named one of the Ten Outstanding Young Scientists by the National Academy for Science and Technology in 2005.
rhechanova@ateneo.edu

Richard Heeks, PhD
One of the founding academics developing the field of ICT4D. Professor of Development Informatics in the Institute for Development Policy and Management, University of Manchester, UK. He is Director of the University's Centre for Development Informatics.
richard.heeks@manchester.ac.uk

Huynh Van Chuong, PhD
Member of the Faculty of Land Resources and Agricultural Environment, Hue University of Agriculture and Forestry. He specialises in land management (land use planning, law and regulations, marketing, non-timber forest products, organisational analysis/development).
huynhvanchuong@huaf.edu.vn

Tahani Iqbal
Former Project Officer and Research Assistant for SIRCA at the Singapore Internet Research Centre, Wee Kim Wee School of Communication and Information, Nanyang Technological University. Prior to this, she was a Research Fellow at *LIRNEasia* where she managed *CPRsouth, LIRNEasia's* capacity and field-building programme to develop an Asia-Pacific knowledge

network on ICT policy regulation; she also conducted research studies on pricing, mobile number portability and other telecom regulatory issues in the Asian region. She is currently an Assistant Manager, Group Regulatory Affairs at Axiata Group Berhad.
tahani.iqbal@gmail.com

May-Ann Lim
Senior Research Analyst for TRP Corporate, a private consulting firm which focuses on telecommunications, media and information communication. Prior to that, she was the manager for Policy Research, Media and Public Education at the Singapore Institute of International Affairs. Her specialities include the development process in Asia, ICT4D, ASEAN CSOs and media studies. She was also part of the World Bank Singapore office, which hosted the 2006 IMF-World Bank Meetings.
mayann@gmail.com

Yvonne Lim
Senior Manager at Singapore Internet Research Centre, Wee Kim Wee School of Communication and Information, Nanyang Technological University. Yvonne has an extensive background in project management; from the Faculty of Arts and Social Sciences (FASS) at the National University of Singapore, she helped the faculty in setting up the Research Clusters in various areas regarded as strategic to FASS' future research development. She has been with the National Institute of Education (NIE), assisting the Head of Academic Group in managing the overall operations of the group.
yvonnelyc@ntu.edu.sg

May O. Lwin, PhD
Associate Professor and Associate Chair at the Wee Kim Wee School of Communication and Information in Nanyang Technological University. She specialises in strategic communication, health and social communication. Dr Lwin has undertaken major cutting-edge research projects in the various areas of strategic communication as well as societal and health communication.
tmaylwin@ntu.edu.sg

Mary Grace P. Mirandilla-Santos
Independent Filipino researcher on ICT and its impact on society, development, and politics. Over the past 10 years, she has conducted research on various topics, including political socialisation, international relations, and public policy. Grace works as a research consultant for the

Centre for Research and Communication, The Asia Foundation, and the Asian Development Bank.
gmirandilla@gmail.com

Ann Mizumoto
Evaluation consultant for the SIRCA programme. She conducted SIRCA's first programme evaluation and subsequently conducted an in-depth study of SIRCA's mentorship scheme. She is passionate about development issues and has done field research and evaluation work with UNICEF (India), UNAIDS (New York), Fondation Scelles (France), the Asia-Pacific Network of People Living with HIV/AIDS (Thailand) and aidha (Singapore). Ann is originally from Sao Paulo, Brazil.
annmizumoto@gmail.com

Pham Huu Ty
Lecturer at the Faculty of Land Resources and Agricultural Environment at Hue University of Agriculture and Forestry (HUAF), Vietnam. He specialises in teaching and research on soil erosion, landslide, land use planning, GIS and remote sensing. He has participated in many rural development projects in the Central Vietnam and very keen to contribute to the protection of severe river landslide and the sustainable livelihoods for river-based relying communities.
phamhuuty@huaf.edu.vn

Chivoin Peou
Lecturer at the Department of Media and Communication, Royal University of Phnom Penh, researcher at the Cambodia Communication Institute, and currently PhD candidate in sociology at the University of Melbourne. His works include articles on Cambodian film and broadcast TV industry, and the annual *Cambodian Communication Review*, published by the Department of Media and Communication.
peouchivoin@gmail.com

Chaitali Sinha
Senior Programme Officer at IDRC. Sinha supports research on how the information society affects people and communities in developing countries, particularly with respect to global health. This includes understanding the role of information and communication technologies (ICTs) in the pursuit of enhanced equity and governance within a strengthened health system. Sinha has worked with IDRC since 2002, supporting applied research on information systems and information society issues in Africa, Asia and Latin

America and the Caribbean. More recently, her work has centred on health information systems and how accurate, relevant, and timely information can improve patient outcomes and public health management. She has a particular interest in the role of open principles (open source, open standards, and open architectures) in health information systems. She has published articles in the field of health information systems and information and communication technologies for development.
csinha@idrc.ca

Matthew Smith, PhD
Programme Officer at IDRC. Oversees research on the use of information and communication technologies (ICTs) to foster sustainable development and social and economic equity; his particular interest is e-government, which is the use of ICTs to improve public sector functioning. Before joining IDRC, Smith did postgraduate research on the interaction between technology and society, in particular the impact of e-government systems on citizens' trust in the government of Chile. Smith also studied e-government in Latin America for the World Bank and the Inter-American Development Bank. He has published on this subject and others, including the concept of openness to broaden access and inclusion.
msmith@idrc.ca

John Traxler, PhD
Professor of Mobile Learning, and Director of Learning Lab at the University of Wolverhampton. He is a Director of the International Association for Mobile Learning, Associate Editor of the International Journal of Mobile and Blended Learning and of Interactive Learning Environments, and is on the Editorial Board of ALT-J and ITID.
John.Traxler@wlv.ac.uk

Santosh Vijaykumar, PhD
PhD (Behavioural Science) from Saint Louis University, USA in 2010. His research broadly focuses on health communication and the use of information and communication technologies (ICTs) in public health. Prior to pursuing doctoral studies, he was involved in international health communication programmes at Johns Hopkins University's, Center for Communication Programmes. His research and commentary can be found in peer-reviewed/ referred publications such as Marriage & Family Review, International Journal of Medical Informatics, Communication Research Reports, Ethnicity & Health, and Economic and Political Weekly.
santoshv@ntu.edu.sg

INDEX